Robot Tales

Eleven Short Science Fiction Stories About Robots

Edited by
Peter I. Kattan

Book Bound Press
https://web.facebook.com/BookboundPress/

Robot Tales
Eleven Short Science Fiction Stories About Robots

Also Available:

Two Short Science Fiction Stories by L.A. Eshbach

Two Short Science Fiction Stories About Mirrors

Ten Stories of Parapsychological Fiction
Short Science Fiction Novels Based on Parapsychology

Dream Stories
Five Short Science Fiction Stories About Dreams

Seven Stories by Philip K. Dick

Timless Stories from Science Fiction Volume 1

Timless Stories from Science Fiction Volume 2

Timless Stories from Science Fiction Volume 3

Contents

Also Available: 2
Arm of the Law 4
The Velvet Glove 29
Survival Tactics 57
Beside Still Waters 72
A Spaceship Named McGuire 80
Service with a Smile 148
The Helpful Robots 154
Robots of the World! Arise! 160
The Love of Frank Nineteen 188
Benefactor 214
The Stutterer 218

Arm of the Law

by Harry Harrison

This story is taken from *Fantastic Universe*, August 1958.

How could a robot—a machine, after all—be involved in something like law application and violence? Harry Harrison, who will be remembered for his THE VELVET GLOVE (Nov. 1956) and his more recent TRAINEE FOR MARS (June 1958) tells what happens when a police robot hits an outpost on Mars.

At one time—this was before the Robot Restriction Laws—they'd even allowed them to make their own decisions....

It was a big, coffin-shaped plywood box that looked like it weighed a ton. This brawny type just dumped it through the door of the police station and started away. I looked up from the blotter and shouted at the trucker's vanishing back.

"What the hell is that?"

"How should I know?" he said as he swung up into the cab. "I just deliver, I don't X-ray 'em. It came on the morning rocket from earth is all I know." He gunned the truck more than he had to and threw up a billowing cloud of red dust.

"Jokers," I growled to myself. "Mars is full of jokers."

When I went over to look at the box I could feel the dust grate between my teeth. Chief Craig must have heard the racket because he came out of his office and helped me stand and look at the box.

"Think it's a bomb?" he asked in a bored voice.

"Why would anyone bother—particularly with a thing this size? And all the way from earth."

He nodded agreement and walked around to look at the other end. There was no sender's address anywhere on the outside. Finally we had to dig out the crowbar and I went to work on the top. After some prying it pulled free and fell off.

That was when we had our first look at Ned. We all would have been a lot happier if it had been our last look as well. If we had just put the lid back on and shipped the thing back to earth! I know now what they mean about Pandora's Box.

But we just stood there and stared like a couple of rubes. Ned lay motionless and stared back at us.

"A robot!" the Chief said.

"Very observant; it's easy to see you went to the police academy."

"Ha ha! Now find out what he's doing here."

I hadn't gone to the academy, but this was no handicap to my finding the letter. It was sticking up out of a thick book in a pocket in the box. The Chief took the letter and read it with little enthusiasm.

"Well, well! United Robotics have the brainstorm that ... *robots, correctly used will tend to prove invaluable in police work* ... they want us to co-operate in a field test ... *robot enclosed is the latest experimental model; valued at 120,000 credits*."

We both looked back at the robot, sharing the wish that the credits had been in the box instead of it. The Chief frowned and moved his lips through the rest of the letter. I wondered how we got the robot out of its plywood coffin.

Experimental model or not, this was a nice-looking hunk of machinery. A uniform navy-blue all over, though the outlet cases, hooks and such were a metallic gold. Someone had gone to a lot of trouble to get that effect. This was as close as a robot could look to a cop in uniform, without being a joke. All that seemed to be missing was the badge and gun.

Then I noticed the tiny glow of light in the robot's eye lenses. It had never occurred to me before that the thing might be turned on. There was nothing to lose by finding out.

"Get out of that box," I said.

The robot came up smooth and fast as a rocket, landing two feet in front of me and whipping out a snappy salute.

"Police Experimental Robot, serial number XPO-456-934B, reporting for duty, sir."

His voice quivered with alertness and I could almost hear the humming of those taut cable muscles. He may have had a stainless steel hide and a bunch of wires for a brain—but he spelled rookie cop to me just the same. The fact that he was man-height with two arms, two legs and that painted-on uniform helped. All I had to do was squint my eyes a bit and there stood Ned the Rookie Cop. Fresh out of school and raring to go. I shook my head to get rid of the illusion. This was just six feet of machine that boffins and brain-boys had turned out for their own amusement.

"Relax, Ned," I said. He was still holding the salute. "At ease. You'll get a hernia of your exhaust pipe if you stay so tense. Anyways, I'm just the sergeant here. That's the Chief of Police over there."

Ned did an about face and slid over to the Chief with that same greased-lightning motion. The Chief just looked at him like something that sprang out from under the hood of a car, while Ned went through the same report routine.

"I wonder if it does anything else beside salute and report," the Chief said while he walked around the robot, looking it over like a dog with a hydrant.

"The functions, operations and responsible courses of action open to the Police Experimental Robots are outlined on pages 184 to 213 of the manual." Ned's voice was muffled for a second while he half-dived back into his case and came up with the volume mentioned. "A detailed breakdown of these will also be found on pages 1035 to 1267 inclusive."

The Chief, who has trouble reading an entire comic page at one sitting, turned the 6-inch-thick book over in his hands like it would maybe bite him. When he had a rough idea of how much it weighed and a good feel of the binding he threw it on my desk.

"Take care of this," he said to me as he headed towards his office. "And the robot, too. Do something with it." The Chief's span of attention never was great and it had been strained to the limit this time.

I flipped through the book, wondering. One thing I never have had much to do with is robots, so I know just as much about them as any Joe in the street. Probably less. The book was filled with pages of fine print, fancy mathematics, wiring diagrams and charts in nine colors and that kind of thing. It needed close attention. Which attention I was not prepared to give at the time. The book slid shut and I eyed the newest employee of the city of Nineport.

"There is a broom behind the door. Do you know how to use it?"

"Yes, sir."

"In that case you will sweep out this room, raising as small a cloud of dust as possible at the same time."

He did a very neat job of it.

I watched 120,000 credits worth of machinery making a tidy pile of butts and sand and wondered why it had been sent to Nineport. Probably because there wasn't another police force in the solar system that was smaller or more unimportant than ours. The engineers must have figured this would be a good spot for a field test. Even if the thing blew up, nobody would really mind. There would probably be someone along some day to get a report on it. Well, they had picked the right spot all right. Nineport was just a little bit beyond nowhere.

Which, of course, was why I was there. I was the only real cop on the force. They needed at least one to give an illusion of the wheels going around. The Chief, Alonzo Craig, had just enough sense to take graft without dropping the money. There were two patrolmen. One old and drunk most of the time. The other so young the only scar he had was the mark of the attram. I had ten years on a metropolitan force, earthside. Why I left is nobody's damn business. I have long since paid for any mistakes I made there by ending up in Nineport.

Nineport is not a city, it's just a place where people stop. The only permanent citizens are the ones who cater to those on the way through. Hotel keepers, restaurant owners, gamblers, barkeeps, and the rest.

There is a spaceport, but only some freighters come there. To pick up the metal from some of the mines that are still working. Some of the settlers still came in for supplies. You might say that Nineport was a town that just missed the boat. In a hundred years I doubt if there will be enough left sticking of the sand to even tell where it used to be. I won't be there either, so I couldn't care less.

I went back to the blotter. Five drunks in the tank, an average night's haul. While I wrote them up Fats dragged in the sixth one.

"Locked himself in the ladies' john at the spaceport and resisting arrest," he reported.

"D and D. Throw him in with the rest."

Fats steered his limp victim across the floor, matching him step for dragging step. I always marveled at the way Fats took care of drunks, since he usually had more under his belt than they had. I have never seen him falling down drunk or completely sober. About all he was good for was keeping a blurred eye on the lockup and running in drunks. He did well at that. No matter what they crawled under or on top of, he found them. No doubt due to the same shared natural instincts.

Fats clanged the door behind number six and weaved his way back in. "What's that?" he asked, peering at the robot along the purple beauty of his nose.

"That is a robot. I have forgotten the number his mother gave him at the factory so we will call him Ned. He works here now."

"Good for him! He can clean up the tank after we throw the bums out."

"That's *my* job," Billy said coming in through the front door. He clutched his nightstick and scowled out from under the brim of his uniform cap. It is not that Billy is stupid, just that most of his strength has gone into his back instead of his mind.

"That's Ned's job now because you have a promotion. You are going to help me with some of my work."

Billy came in very handy at times and I was anxious that the force shouldn't lose him. My explanation cheered him because he sat down by Fats and watched Ned do the floor.

That's the way things went for about a week. We watched Ned sweep and polish until the station began to take on a positively antiseptic look. The Chief, who always has an eye out for that type of thing, found out that Ned could file the odd ton of reports and paperwork that cluttered his office. All this kept the robot busy, and we got so used to him we were hardly aware he was around. I knew he had moved the packing case into the storeroom and fixed himself up a cozy sort of robot dormitory-coffin. Other than that I didn't know or care.

The operation manual was buried in my desk and I never looked at it. If I had, I might have had some idea of the big changes that were in store. None of us knew the littlest bit about what a robot can or cannot do. Ned was working nicely as a combination janitor-file clerk and should have stayed that way. He would have too if the Chief hadn't been so lazy. That's what started it all.

It was around nine at night and the Chief was just going home when the call came in. He took it, listened for a moment, then hung up.

"Greenback's liquor store. He got held up again. Says to come at once."

"That's a change. Usually we don't hear about it until a month later. What's he paying protection money for if China Joe ain't protecting? What's the rush now?"

The Chief chewed his loose lip for a while, finally and painfully reached a decision.

"You better go around and see what the trouble is."

"Sure," I said reaching for my cap. "But no one else is around, you'll have to watch the desk until I get back."

"That's no good," he moaned. "I'm dying from hunger and sitting here isn't going to help me any."

"I will go take the report," Ned said, stepping forward and snapping his usual well-greased salute.

At first the Chief wasn't buying. You would think the water cooler came to life and offered to take over his job.

"How could *you* take a report?" he growled, putting the wise-guy water cooler in its place. But he had phrased his little insult as a question so he had only himself to blame. In exactly three minutes Ned gave the Chief a summary of the routine necessary for a police officer to make a report on an armed robbery or other reported theft. From the glazed look in Chief's protruding eyes I could tell Ned had quickly passed the boundaries of the Chief's meager knowledge.

"Enough!" the harried man finally gasped. "If you know so much why don't you make a report?"

Which to me sounded like another version of "*if you're so damned smart why ain't you rich?*" which we used to snarl at the brainy kids in grammar school. Ned took such things literally though, and turned towards the door.

"Do you mean you wish me to make a report on this robbery?"

"Yes," the Chief said just to get rid of him, and we watched his blue shape vanish through the door.

"He must be brighter than he looks," I said. "He never stopped to ask where Greenback's store is."

The Chief nodded and the phone rang again. His hand was still resting on it so he picked it up by reflex. He listened for a second and you would have thought someone was pumping blood out of his heel from the way his face turned white.

"The holdup's still on," he finally gasped. "Greenback's delivery boy is on the line—calling back to see where we are. Says he's under a table in the back room ..."

I never heard the rest of it because I was out the door and into the car. There were a hundred things that could happen if Ned got there before me. Guns could go off, people hurt, lots of things. And the police would be to blame for it all—sending a tin robot to do a cop's job. Maybe the Chief had ordered Ned there, but clearly as if the words were painted on the windshield of the car, I knew I would be dragged into it. It never gets very warm on Mars, but I was sweating.

Nineport has fourteen traffic regulations and I broke all of them before I had gone a block. Fast as I was, Ned was faster. As I turned the corner I saw him open the door of Greenback's store and walk in. I screamed brakes in behind him and arrived just in time to have a gallery seat. A shooting gallery at that.

There were two holdup punks, one behind the counter making like a clerk and the other lounging off to the side. Their guns were out of sight, but blue-coated Ned busting through the door like that was too much for their keyed up nerves. Up came both guns like they were on strings and Ned stopped dead. I grabbed for my own gun and waited for pieces of busted robot to come flying through the window.

Ned's reflexes were great. Which I suppose is what you should expect of a robot.

"DROP YOUR GUNS, YOU ARE UNDER ARREST."

He must have had on full power or something, his voice blasted so loud my ears hurt. The result was just what you might expect. Both torpedoes let go at once and the air was filled with flying slugs. The show windows went out with a crash and I went down on my stomach. From the amount of noise I knew they both had recoilless .50's. You can't stop one of those slugs. They go right through you and anything else that happens to be in the way.

Except they didn't seem to be bothering Ned. The only notice he seemed to take was to cover his eyes. A little shield with a thin slit popped down over his eye lenses. Then he moved in on the first thug.

I knew he was fast, but not that fast. A couple of slugs jarred him as he came across the room, but before the punk could change his aim Ned had the gun in his hand. That was the end of that. He put on one of the sweetest hammer locks I have ever seen and neatly grabbed the gun when it dropped from the limp fingers. With the same motion that slipped the gun into a pouch he whipped out a pair of handcuffs and snapped them on the punk's wrists.

Holdupnik number two was heading for the door by then, and I was waiting to give him a warm reception. There was never any need. He hadn't gone halfway before Ned slid in front of him. There was a thud when they hit that didn't even shake Ned, but gave the other a glazed look. He never even knew it when Ned slipped the cuffs on him and dropped him down next to his partner.

I went in, took their guns from Ned, and made the arrest official. That was all Greenback saw when he crawled out from behind the counter and it was all I wanted him to see. The place was a foot deep in broken glass and smelled like the inside of a Jack Daniels bottle. Greenback began to howl like a wolf over his lost stock. He didn't seem to know any more about the phone call than I did, so I grabbed ahold of a pimply looking kid who staggered out of the storeroom. He was the one who had made the calls.

It turned out to be a matter of sheer stupidity. He had worked for Greenback only a few days and didn't have enough brains to realize that all holdups should be reported to the protection boys instead of the police. I told Greenback to wise up his boy, as look at the trouble that got caused. Then pushed the two ex-holdup men out to the car. Ned climbed in back with them and they clung together like two waifs in a storm. The robot's only response was to pull a first aid kit from his hip and fix up a ricochet hole in one of the thugs that no one had noticed in the excitement.

The Chief was still sitting there with that bloodless look when we marched in. I didn't believe it could be done, but he went two shades whiter.

"You made the pinch," he whispered. Before I could straighten him out a second and more awful idea hit him. He grabbed a handful of shirt on the first torpedo and poked his face down. "You with China Joe," he snarled.

The punk made the error of trying to be cute so the Chief let him have one on the head with the open hand that set his eyes rolling like marbles. When the question got asked again he found the right answer.

"I never heard from no China Joe. We just hit town today and—"

"Freelance, by God," the Chief sighed and collapsed into his chair. "Lock 'em up and quickly tell me what in hell happened."

I slammed the gate on them and pointed a none too steady finger at Ned.

"There's the hero," I said. "Took them on single-handed, rassled them for a fall and made the capture. He is a one-robot tornado, a power for good in this otherwise evil community. And he's bulletproof too." I ran a finger over Ned's broad chest. The paint was chipped by the slugs, but the metal was hardly scratched.

"This is going to cause me trouble, big trouble," the Chief wailed.

I knew he meant with the protection boys. They did not like punks getting arrested and guns going off without their okay. But Ned thought the Chief had other worries and rushed in to put them right. "There will be no trouble. At no time did I violate any of the Robotic Restriction Laws, they are part of my control circuits and therefore fully automatic. The men who drew their guns violated both robotic and human law when they threatened violence. I did not injure the men—merely restrained them."

It was all over the Chief's head, but I liked to think *I* could follow it. And I *had* been wondering how a robot—a machine—could be involved in something like law application and violence. Ned had the answer to that one too.

"Robots have been assuming these functions for years. Don't recording radar meters pass judgment on human violation of automobile regulations? A robot alcohol detector is better qualified to assess the sobriety of a prisoner than the arresting officer. At one time robots were even allowed to make their own decisions about killing. Before the Robotic Restriction Laws automatic gun-pointers were in general use. Their final development was a self-contained battery of large anti-aircraft guns. Automatic scan radar detected all aircraft in the vicinity. Those that could not return the correct identifying signal had their courses tracked and computed, automatic fuse-cutters and loaders readied the computer-aimed guns—which were fired by the robot mechanism."

There was little I could argue about with Ned. Except maybe his college-professor vocabulary. So I switched the attack.

"But a robot can't take the place of a cop, it's a complex human job."

"Of course it is, but taking a human policeman's place is not the function of a police robot. Primarily I combine the functions of numerous pieces of police equipment, integrating their operations and making them instantly available. In addition I can aid in the *mechanical* processes of law enforcement. If you arrest a man you handcuff him. But if you order me to do it, I have made no moral decision. I am just a machine for attaching handcuffs at that point ..."

My raised hand cut off the flow of robotic argument. Ned was hipped to his ears with facts and figures and I had a good idea who would come off second best in any continued discussion. No laws had been broken when Ned made the pinch, that was for sure. But there are other laws than those that appear on the books.

"China Joe is not going to like this, not at all," the Chief said, speaking my own thoughts.

The law of Tooth and Claw. That's one that wasn't in the law books. And that was what ran Nineport. The place was just big enough to have a good population of gambling joints, bawdy houses and drunk-rollers. They were all run by China Joe. As was the police department. We were all in his pocket and you might say he was the one who paid our wages. This is not the kind of thing, though, that you explain to a robot.

"Yeah, China Joe."

I thought it was an echo at first, then realized that someone had eased in the door behind me. Something called Alex. Six feet of bone, muscle and trouble. China Joe's right hand man. He imitated a smile at the Chief who sank a bit lower in his chair.

"China Joe wants you should tell him why you got smart cops going around and putting the arm on people and letting them shoot up good liquor. He's mostly angry about the hooch. He says that he had enough guff and after this you should—"

"I am putting you under Robot Arrest, pursuant to article 46, paragraph 19 of the revised statutes ..."

Ned had done it before we realized he had even moved. Right in front of our eyes he was arresting Alex and signing our death warrants.

Alex was not slow. As he turned to see who had grabbed him, he had already dragged out this cannon. He got one shot in, square against Ned's chest, before the robot plucked the gun away and slipped on the cuffs. While we all gaped like dead fish, Ned recited the charge in what I swear was a satisfied tone.

"The prisoner is Peter Rakjomskj, alias Alex the Axe, wanted in Canal City for armed robbery and attempted murder. Also wanted by local police of Detroit, New York and Manchester on charges of ..."

"*Get it off me!*" Alex howled. We might have too, and everything might have still been straightened out if Benny Bug hadn't heard the shot. He popped his head in the front door just long enough to roll his eyes over our little scene.

"Alex ... they're puttin' the arm on Alex!"

Then he was gone and when I hit the door he was nowhere in sight. China Joe's boys always went around in pairs. And in ten minutes he would know all about it.

"Book him," I told Ned. "It wouldn't make any difference if we let him go now. The world has already come to an end."

Fats came in then, mumbling to himself. He jerked a thumb over his shoulder when he saw me.

"What's up? I see little Benny Bug come out of here like the place was on fire and almost get killed driving away?"

Then Fats saw Alex with the bracelets on and turned sober in one second. He just took a moment to gape, then his mind was made up. Without a trace of a stagger he walked over to the Chief and threw his badge on the desk in front of him.

"I am an old man and I drink too much to be a cop. Therefore I am resigning from the force. Because if that is whom I think it is over there with the cuffs on, I will not live to be a day older as long as I am around here."

"Rat." The Chief growled in pain through his clenched teeth. "Deserting the sinking ship. Rat."

"Squeak," Fats said and left.

The Chief was beyond caring at this point. He didn't blink an eye when I took Fats' badge off the desk. I don't know why I did it, perhaps I thought it was only fair. Ned had started all the trouble and I was just angry enough to want him on the spot when it was finished. There were two rings on his chest plate, and I was not surprised when the badge pin fitted them neatly.

"There, now you are a real cop." Sarcasm dripped from the words. I should have realized that robots are immune to sarcasm. Ned took my statement at face value.

"This is a very great honor, not only for me but for all robots. I will do my best to fulfill all the obligations of the office." Jack Armstrong in tin underwear. I could hear the little motors in his guts humming with joy as he booked Alex.

If everything else hadn't been so bad I would have enjoyed that. Ned had more police equipment built into him than

Nineport had ever owned. There was an ink pad that snapped out of one hip, and he efficiently rolled Alex's fingertips across it and stamped them on a card. Then he held the prisoner at arm's length while something clicked in his abdomen. Once more sideways and two instant photographs dropped out of a slot. The mug shots were stuck on the card, arrest details and such inserted. There was more like this, but I forced myself away. There were more important things to think about.

Like staying alive.

"Any ideas, Chief?"

A groan was my only answer so I let it go at that. Billy, the balance of the police force, came in then. I gave him a quick rundown. Either through stupidity or guts he elected to stay, and I was proud of the boy. Ned locked away the latest prisoner and began sweeping up.

That was the way we were when China Joe walked in.

Even though we were expecting it, it was still a shock. He had a bunch of his toughest hoods with him and they crowded through the door like an overweight baseball team. China Joe was in front, hands buried in the sleeves of his long mandarin gown. No expression at all on his ascetic features. He didn't waste time talking to us, just gave the word to his own boys.

"Clean this place up. The new police Chief will be here in a while and I don't want him to see any bums hanging around."

It made me angry. Even with the graft I like to feel I'm still a cop. Not on a cheap punk's payroll. I was also curious about China Joe. Had been ever since I tried to get a line on him and never found a thing. I still wanted to know.

"Ned, take a good look at that Chinese guy in the rayon bathrobe and let me know who he is."

My, but those electronic circuits work fast. Ned shot the answer back like a straight man who had been rehearsing his lines for weeks.

"He is a pseudo-oriental, utilizing a natural sallowness of the skin heightened with dye. He is not Chinese. There has also been an operation on his eyes, scars of which are still visible. This has been undoubtedly done in an attempt to conceal his real identity, but Bertillon measurements of his ears and other features make identity positive. He is on the Very Wanted list of Interpol and his real name is ..."

China Joe was angry, and with a reason.

"That's the *thing* ... that big-mouthed tin radio set over there. We heard about it and we're taking care of it!"

The mob jumped aside then or hit the deck and I saw there was a guy kneeling in the door with a rocket launcher. Shaped anti-tank charges, no doubt. That was my last thought as the thing let go with a "whoosh."

Maybe you can hit a tank with one of those. But not a robot. At least not a police robot. Ned was sliding across the floor on his face when the back wall blew up. There was no second shot. Ned closed his hand on the tube of the bazooka and it was so much old drainpipe.

Billy decided then that anyone who fired a rocket in a police station was breaking the law, so he moved in with his club. I was right behind him since I did not want to miss any of the fun. Ned was at the bottom somewhere, but I didn't doubt he could take care of himself.

There were a couple of muffled shots and someone screamed. No one fired after that because we were too tangled up. A punk named Brooklyn Eddie hit me on the side of the head with his gunbutt and I broke his nose all over his face with my fist.

There is a kind of a fog over everything after that. But I do remember it was very busy for a while.

When the fog lifted a bit I realized I was the only one still standing. Or leaning rather. It was a good thing the wall was there.

Ned came in through the street door carrying a very bashed-looking Brooklyn Eddie. I hoped I had done all that. Eddie's wrists were fastened together with cuffs. Ned laid him gently next to the heap of thugs—who I suddenly realized all wore the same kind of handcuffs. I wondered vaguely if Ned made them as he needed them or had a supply tucked away in a hollow leg or something.

There was a chair a few feet away and sitting down helped.

Blood was all over everything and if a couple of the hoods hadn't groaned I would have thought they were corpses. One was, I noticed suddenly. A bullet had caught him in the chest, most of the blood was probably his.

Ned burrowed in the bodies for a moment and dragged Billy out. He was unconscious. A big smile on his face and the splintered remains of his nightstick still stuck in his fist. It takes very little to make some people happy. A bullet had gone through his leg and he never moved while Ned ripped the pants leg off and put on a bandage.

"The spurious China Joe and one other man escaped in a car," Ned reported.

"Don't let it worry you," I managed to croak. "Your batting average still leads the league."

It was then I realized the Chief was still sitting in his chair, where he had been when the brouhaha started. Still slumped down with that glazed look. Only after I started to talk to him did I realize that Alonzo Craig, Chief of Police of Nineport, was now dead.

A single shot. Small caliber gun, maybe a .22. Right through the heart and what blood there had been was soaked up by his clothes. I had a good idea where the gun would be that fired that shot. A small gun, the kind that would fit in a wide Chinese sleeve.

I wasn't tired or groggy any more. Just angry. Maybe he hadn't been the brightest or most honest guy in the world. But he deserved a better end than that. Knocked off by a two-bit racket boss who thought he was being crossed.

Right about then I realized I had a big decision to make. With Billy out of the fight and Fats gone I was the Nineport police force. All I had to do to be clear of this mess was to walk out the door and keep going. I would be safe enough.

Ned buzzed by, picked up two of the thugs, and hauled them off to the cells.

Maybe it was the sight of his blue back or maybe I was tired of running. Either way my mind was made up before I realized it. I carefully took off the Chief's gold badge and put it on in place of my old one.

"The new Chief of Police of Nineport," I said to no one in particular.

"Yes, sir," Ned said as he passed. He put one of the prisoners down long enough to salute, then went on with his work. I returned the salute.

The hospital meat wagon hauled away the dead and wounded. I took an evil pleasure in ignoring the questioning stares of the attendants. After the doc fixed the side of my head, everyone cleared out. Ned mopped up the floor. I ate ten aspirin and waited for the hammering to stop so I could think what to do next.

When I pulled my thoughts together the answer was obvious. Too obvious. I made as long a job as I could of reloading my gun.

"Refill your handcuff box, Ned. We are going out."

Like a good cop he asked no questions. I locked the outside door when we left and gave him the key.

"Here. There's a good chance you will be the only one left to use this before the day is over."

I stretched the drive over to China Joe's place just as much as I could. Trying to figure if there was another way of doing it. There wasn't. Murder had been done and Joe was the boy I was going to pin it on. So I had to get him.

The best I could do was stop around the corner and give Ned a briefing.

"This combination bar and dice-room is the sole property of he whom we will still call China Joe until there is time for you to give me a rundown on him. Right now I got enough distractions. What we have to do is go in there, find Joe and bring him to justice. Simple?"

"Simple," Ned answered in his sharp Joe-college voice. "But wouldn't it be simpler to make the arrest now, when he is leaving in that car, instead of waiting until he returns?"

The car in mention was doing sixty as it came out of the alley ahead of us. I only had a glimpse of Joe in the back seat as it tore by us.

"Stop them!" I shouted, mostly for my own benefit since I was driving. I tried to shift gears and start the engine at the same time, and succeeded in doing exactly nothing.

So Ned stopped them. It had been phrased as an order. He leaned his head out of the window and I saw at once why most of his equipment was located in his torso. Probably his brain as well. There sure wasn't much room left in his head when that cannon was tucked away in there.

A .75 recoilless. A plate swiveled back right where his nose should have been if he had one, and the big muzzle pointed out. It's a neat idea when you think about it. Right between the eyes for good aiming, up high, always ready.

The BOOM BOOM almost took my head off. Of course Ned was a perfect shot—so would I be with a computer for a brain. He had holed one rear tire with each slug and the car flap-flapped to a stop a little ways down the road. I climbed out slowly while Ned sprinted there in seconds flat. They didn't even try to run this time. What little nerve they had left must have been shattered by the smoking muzzle of that .75 poking out from between Ned's eyes. Robots are neat about things like that so he must have left it sticking out deliberate. Probably had a course in psychology back in robot school.

Three of them in the car, all waving their hands in the air like the last reel of a western. And the rear floor covered with interesting little suitcases.

Everyone came along quietly.

China Joe only snarled while Ned told me that his name really was Stantin and the Elmira hot seat was kept warm all the time in hopes he would be back. I promised Joe-Stantin I would be happy to arrange it that same day. Thereby not worrying about any slip-ups with the local authorities. The rest of the mob would stand trial in Canal City.

It was a very busy day.

Things have quieted down a good deal since then. Billy is out of the hospital and wearing my old sergeant's stripes. Even Fats is back, though he is sober once in a while now and has trouble looking me in the eye. We don't have much to do because in addition to being a quiet town this is now an honest one.

Ned is on foot patrol nights and in charge of the lab and files days. Maybe the Policeman's Benevolent wouldn't like that, but Ned doesn't seem to mind. He touched up all the bullet scratches and keeps his badge polished. I know a robot can't be happy or sad—but Ned *seems* to be happy.

Sometimes I would swear I can hear him humming to himself. But, of course, that is only the motors and things going around.

When you start thinking about it, I suppose we set some kind of precedent here. What with putting on a robot as a full-fledged police officer. No one ever came around from the factory yet, so I have never found out if we're the first or not.

And I'll tell you something else. I'm not going to stay in this broken-down town forever. I have some letters out now, looking for a new job.

So some people are going to be *very* surprised when they see who their new Chief of Police is after *I* leave.

The Velvet Glove

by Harry Harrison

This story is taken from *Fantastic Universe,* November 1956.

SF writer and editor Harry Harrison explores a not too distant future where robots—particularly specialist robots who don't know their place—have quite a rough time of it. True, the Robot Equality Act had been passed—but so what?

New York was a bad town for robots this year. In fact, all over the country it was bad for robots....

Jon Venex fitted the key into the hotel room door. He had asked for a large room, the largest in the hotel, and paid the desk clerk extra for it. All he could do now was pray that he hadn't been cheated. He didn't dare complain or try to get his money back. He heaved a sigh of relief as the door swung open, it was bigger than he had expected—fully three feet wide by five feet long. There was more than enough room to work in. He would have his leg off in a jiffy and by morning his limp would be gone.

There was the usual adjustable hook on the back wall. He slipped it through the recessed ring in the back of his neck and kicked himself up until his feet hung free of the floor. His legs relaxed with a rattle as he cut off all power from his waist down.

The overworked leg motor would have to cool down before he could work on it, plenty of time to skim through the newspaper. With the chronic worry of the unemployed, he snapped it open at the want-ads and ran his eye down the

Help Wanted—Robot column. There was nothing for him under the Specialist heading, even the Unskilled Labor listings were bare and unpromising. New York was a bad town for robots this year.

The want-ads were just as depressing as usual but he could always get a lift from the comic section. He even had a favorite strip, a fact that he scarcely dared mention to himself. "Rattly Robot," a dull-witted mechanical clod who was continually falling over himself and getting into trouble. It was a repellent caricature, but could still be very funny. Jon was just starting to read it when the ceiling light went out.

It was ten P.M., curfew hour for robots. Lights out and lock yourself in until six in the morning, eight hours of boredom and darkness for all except the few night workers. But there were ways of getting around the letter of a law that didn't concern itself with a definition of visible light. Sliding aside some of the shielding around his atomic generator, Jon turned up the gain. As it began to run a little hot the heat waves streamed out—visible to him as infra-red rays. He finished reading the paper in the warm, clear light of his abdomen.

The thermocouple in the tip of his second finger left hand, he tested the temperature of his leg. It was soon cool enough to work on. The waterproof gasket stripped off easily, exposing the power leads, nerve wires and the weakened knee joint. The wires disconnected, Jon unscrewed the knee above the joint and carefully placed it on the shelf in front of him. With loving care he took the replacement part from his hip pouch. It was the product of toil, purchased with his savings from three months employment on the Jersey pig farm.

Jon was standing on one leg testing the new knee joint when the ceiling fluorescent flickered and came back on. Five-thirty already, he had just finished in time. A shot of oil on the new bearing completed the job; he stowed away the tools in the pouch and unlocked the door.

The unused elevator shaft acted as waste chute, he slipped his newspaper through a slot in the door as he went by. Keeping close to the wall, he picked his way carefully down the grease-stained stairs. He slowed his pace at the 17th floor as two other mechs turned in ahead of him. They were obviously butchers or meat-cutters; where the right hand should have been on each of them there stuck out a wicked, foot-long knife. As they approached the foot of the stairs they stopped to slip the knives into the plastic sheaths that were bolted to their chestplates. Jon followed them down the ramp into the lobby.

The room was filled to capacity with robots of all sizes, forms and colors. Jon Venex's greater height enabled him to see over their heads to the glass doors that opened onto the street. It had rained the night before and the rising sun drove red glints from the puddles on the sidewalk. Three robots, painted snow white to show they were night workers, pushed the doors open and came in. No one went out as the curfew hadn't ended yet. They milled around slowly talking in low voices.

The only human being in the entire lobby was the night clerk dozing behind the counter. The clock over his head said five minutes to six. Shifting his glance from the clock, Jon became aware of a squat black robot waving to attract his attention. The powerful arms and compact build identified him as a member of the Diger family, one of the most numerous groups. He pushed through the crowd and clapped Jon on the back with a resounding clang.

"Jon Venex! I knew it was you as soon as I saw you sticking up out of this crowd like a green tree trunk. I haven't seen you since the old days on Venus!"

Jon didn't need to check the number stamped on the short one's scratched chestplate. Alec Diger had been his only close friend during those thirteen boring years at Orange Sea Camp. A good chess player and a whiz at Two-handed Handball, they had spent all their off time together. They shook hands, with the extra squeeze that means friendliness.

"Alec, you beat-up little grease pot, what brings you to New York?"

"The burning desire to see something besides rain and jungle, if you must know. After you bought out, things got just too damn dull. I began working two shifts a day in that foul diamond mine, and then three a day for the last month to get enough credits to buy my contract and passage back to earth. I was underground so long that the photocell on my right eye burned out when the sunlight hit it."

He leaned forward with a hoarse confidential whisper, "If you want to know the truth, I had a sixty-carat diamond stuck behind the eye lens. I sold it here on earth for two hundred credits, gave me six months of easy living. It's all gone now, so I'm on my way to the employment exchange." His voice boomed loud again, "And how about *you*?"

Jon Venex chuckled at his friend's frank approach to life. "It's just been the old routine with me, a run of odd jobs until I got side-swiped by a bus—it fractured my knee bearing. The only job I could get with a bad leg was feeding slops to pigs. Earned enough to fix the knee—and here *I* am."

Alec jerked his thumb at a rust-colored, three-foot-tall robot that had come up quietly beside him. "If you think you've got trouble take a look at Dik here, that's no coat of paint on him. Dik Dryer, meet Jon Venex an old buddy of mine."

Jon bent over to shake the little mech's hand. His eye shutters dilated as he realized what he had thought was a coat of paint was a thin layer of rust that coated Dik's metal body. Alec scratched a shiny path in the rust with his fingertip. His voice was suddenly serious.

"Dik was designed for operation in the Martian desert. It's as dry as a fossil bone there so his skinflint company cut corners on the stainless steel.

"When they went bankrupt he was sold to a firm here in the city. After a while the rust started to eat in and slow him down, they gave Dik his contract and threw him out."

The small robot spoke for the first time, his voice grated and scratched. "Nobody will hire me like this, but I can't get repaired until I get a job." His arms squeaked and grated as he moved them. "I'm going by the Robot Free Clinic again today, they said they might be able to do something."

Alec Diger rumbled in his deep chest. "Don't put too much faith in those people. They're great at giving out tenth-credit oil capsules or a little free wire—but don't depend on them for anything important."

It was six now, the robots were pushing through the doors into the silent streets. They joined the crowd moving out, Jon slowing his stride so his shorter friends could keep pace. Dik Dryer moved with a jerking, irregular motion, his voice as uneven as the motion of his body.

"Jon—Venex, I don't recognize your family name. Something to do—with Venus—perhaps."

"Venus is right, Venus Experimental—there are only twenty-two of us in the family. We have waterproof, pressure-resistant bodies for working down on the ocean bottom. The basic idea was all right, we did our part, only there wasn't enough money in the channel-dredging contract to keep us all working. I bought out my original contract at half price and became a free robot."

Dik vibrated his rusted diaphragm. "Being free isn't all it should be. I some—times wish the Robot Equality Act hadn't been passed. I would just l-love to be owned by a nice rich company with a machine shop and a—mountain of replacement parts."

"You don't really mean that, Dik," Alec Diger clamped a heavy black arm across his shoulders. "Things aren't perfect now, we know that, but it's certainly a lot better than the old days, we were just hunks of machinery then. Used twenty-four hours a day until we were worn out and then thrown in the junk pile. No thanks, I'll take my chances with things as they are."

Jon and Alec turned into the employment exchange, saying good-by to Dik who went on slowly down the street. They pushed up the crowded ramp and joined the line in front of the registration desk. The bulletin board next to the desk held a scattering of white slips announcing job openings. A clerk was pinning up new additions.

Venex scanned them with his eyes, stopping at one circled in red.

ROBOTS NEEDED IN THESE CATEGORIES. APPLY AT ONCE TO CHAINJET, LTD., 1219 BROADWAY.

Fasten
Flyer
Atommel
Filmer
Venex

Jon rapped excitedly on Alec Diger's neck. "Look there, a job in my own specialty—I can get my old pay rate! See you back at the hotel tonight—and good luck in your job hunting."

Alec waved good-by. "Let's hope the job's as good as you think, I never trust those things until I have my credits in my hand."

Jon walked quickly from the employment exchange, his long legs eating up the blocks. *Good old Alec, he didn't believe in anything he couldn't touch. Perhaps he was right, but why try to be unhappy. The world wasn't too bad this morning—his leg worked fine, prospects of a good job—he hadn't felt this cheerful since the day he was activated.*

Turning the corner at a brisk pace he collided with a man coming from the opposite direction. Jon had stopped on the instant, but there wasn't time to jump aside. The obese individual jarred against him and fell to the ground. From the height of elation to the depths of despair in an instant—he had injured a *human being*!

He bent to help the man to his feet, but the other would have none of that. He evaded the friendly hand and screeched in a high-pitched voice.

"Officer, officer, police ... HELP! I've been attacked—a mad robot ... HELP!"

A crowd was gathering—staying at a respectful distance—but making an angry muttering noise. Jon stood motionless, his head reeling at the enormity of what he had done. A policeman pushed his way through the crowd.

"Seize him, officer, shoot him down ... he struck me ... almost killed me ..." The man shook with rage, his words thickening to a senseless babble.

The policeman had his .75 recoilless revolver out and pressed against Jon's side.

"This *man* has charged you with a serious crime, *grease-can*. I'm taking you into the station house—to talk about it." He looked around nervously, waving his gun to open a path through the tightly packed crowd. They moved back grudgingly, with murmurs of disapproval.

Jon's thoughts swirled in tight circles. How did a catastrophe like this happen, where was it going to end? He didn't dare tell the truth, that would mean he was calling the man a liar. There had been six robots power-lined in the city since the first of the year. If he dared speak in his own defense there would be a jumper to the street lighting circuit and a seventh burnt out hulk in the police morgue.

A feeling of resignation swept through him, there was no way out. If the man pressed charges it would mean a term of penal servitude, though it looked now as if he would never live to reach the court. The papers had been whipping up a lot of anti-robe feeling, you could feel it behind the angry voices, see it in the narrowed eyes and clenched fists. The crowd was slowly changing into a mob, a mindless mob as yet, but capable of turning on him at any moment.

"What's goin' on here...?" It was a booming voice, with a quality that dragged at the attention of the crowd.

A giant cross-continent freighter was parked at the curb. The driver swung down from the cab and pushed his way through the people. The policeman shifted his gun as the man strode up to him.

"That's my robot you got there, Jack, don't put any holes in him!" He turned on the man who had been shouting accusations. "Fatty here, is the world's biggest liar. The robot was standing here waiting for me to park the truck. Fatty must be as blind as he is stupid, I saw the whole thing. He knocks himself down walking into the robe, then starts hollering for the cops."

The other man could take no more. His face crimson with anger he rushed toward the trucker, his fists swinging in ungainly circles. They never landed, the truck driver put a meaty hand on the other's face and seated him on the sidewalk for the second time.

The onlookers roared with laughter, the power-lining and the robot were forgotten. The fight was between two men now, the original cause had slipped from their minds. Even the policeman allowed himself a small smile as he holstered his gun and stepped forward to separate the men.

The trucker turned towards Jon with a scowl.

"Come on you aboard the truck—you've caused me enough trouble for one day. What a junkcan!"

The crowd chuckled as he pushed Jon ahead of him into the truck and slammed the door behind them. Jamming the starter with his thumb he gunned the thunderous diesels into life and pulled out into the traffic.

Jon moved his jaw, but there were no words to come out. Why had this total stranger helped him, what could he say to show his appreciation? He knew that all humans weren't robe-haters, why it was even rumored that some humans treated robots as *equals* instead of machines. The driver must be one of these mythical individuals, there was no other way to explain his actions.

Driving carefully with one hand the man reached up behind the dash and drew out a thin, plastikoid booklet. He handed it to Jon who quickly scanned the title, *Robot Slaves in a World Economy* by Philpott Asimov II.

"If you're caught reading that thing they'll execute you on the spot. Better stick it between the insulation on your generator, you can always burn it if you're picked up.

"Read it when you're alone, it's got a lot of things in it that you know nothing about. Robots aren't really inferior to humans, in fact they're superior in most things. There is even a little history in there to show that robots aren't the first ones to be treated as second class citizens. You may find it a little hard to believe, but human beings once treated each other just the way they treat robots now. That's one of the reasons I'm active in this movement—sort of like the fellow who was burned helping others stay away from the fire."

He smiled a warm, friendly smile in Jon's direction, the whiteness of his teeth standing out against the rich ebony brown of his features.

"I'm heading towards US-1, can I drop you anywheres on the way?"

"The Chainjet Building please—I'm applying for a job."

They rode the rest of the way in silence. Before he opened the door the driver shook hands with Jon.

"Sorry about calling you *junkcan*, but the crowd expected it." He didn't look back as he drove away.

Jon had to wait a half hour for his turn, but the receptionist finally signalled him towards the door of the interviewer's room. He stepped in quickly and turned to face the man seated at the transplastic desk, an upset little man with permanent worry wrinkles stamped in his forehead. The little man shoved the papers on the desk around angrily, occasionally making crabbed little notes on the margins. He flashed a birdlike glance up at Jon.

"Yes, yes, be quick. What is it you want?"

"You posted a help wanted notice, I—"

The man cut him off with a wave of his hand. "All right let me see your ID tag ... quickly, there are others waiting."

Jon thumbed the tag out of his waist slot and handed it across the desk. The interviewer read the code number, then began running his finger down a long list of similar figures. He stopped suddenly and looked sideways at Jon from under his lowered lids.

"You have made a mistake, we have no opening for you."

Jon began to explain to the man that the notice had requested his specialty, but he was waved to silence. As the interviewer handed back the tag he slipped a card out from under the desk blotter and held it in front of Jon's eyes. He held it there for only an instant, knowing that the written message was recorded instantly by the robot's photographic vision and eidetic memory. The card dropped into the ash tray and flared into embers at the touch of the man's pencil-heater.

Jon stuffed the ID tag back into the slot and read over the message on the card as he walked down the stairs to the street. There were six lines of typewritten copy with no signature.

To Venex Robot: You are urgently needed on a top secret company project. There are suspected informers in the main office, so you are being hired in this unusual manner. Go at once to 787 Washington Street and ask for Mr. Coleman.

Jon felt an immense sensation of relief. For a moment there, he was sure the job had been a false lead. He saw nothing unusual in the method of hiring. The big corporations were immensely jealous of their research discoveries and went to great lengths to keep them secret—at the same time resorting to any means to ferret out their business rivals' secrets. There might still be a chance to get this job.

The burly bulk of a lifter was moving back and forth in the gloom of the ancient warehouse stacking crates in ceiling-high rows. Jon called to him, the robot swung up his forklift and rolled over on noiseless tires. When Jon questioned him he indicated a stairwell against the rear wall.

"Mr. Coleman's office is down in back, the door is marked." The lifter put his fingertips against Jon's ear pick-ups and lowered his voice to the merest shadow of a whisper. It would have been inaudible to human ears, but Jon could hear him easily, the sounds being carried through the metal of the other's body.

"He's the meanest man you ever met—he hates robots so be *ever* so polite. If you can use 'sir' five times in one sentence you're perfectly safe."

Jon swept the shutter over one eye tube in a conspiratorial wink, the large mech did the same as he rolled away. Jon turned and went down the dusty stairwell and knocked gently on Mr. Coleman's door.

Coleman was a plump little individual in a conservative purple-and-yellow business suit. He kept glancing from Jon to the Robot General Catalog checking the Venex specifications listed there. Seemingly satisfied he slammed the book shut.

"Gimme your tag and back against that wall to get measured."

Jon laid his ID tag on the desk and stepped towards the wall. "Yes, sir, here it is, sir." Two "sir" on that one, not bad for the first sentence. He wondered idly if he could put five of them in one sentence without the man knowing he was being made a fool of.

He became aware of the danger an instant too late. The current surged through the powerful electromagnet behind the plaster flattening his metal body helplessly against the wall. Coleman was almost dancing with glee.

"We got him, Druce, he's mashed flatter than a stinking tin-can on a rock, can't move a motor. Bring that junk in here and let's get him ready."

Druce had a mechanic's coveralls on over his street suit and a tool box slung under one arm. He carried a little black metal can at arm's length, trying to get as far from it as possible. Coleman shouted at him with annoyance.

"That bomb can't go off until it's armed, stop acting like a child. Put it on that grease-can's leg and *quick*!"

Grumbling under his breath, Druce spot-welded the metal flanges of the bomb onto Jon's leg a few inches above his knee. Coleman tugged at it to be certain it was secure, then twisted a knob in the side and pulled out a glistening length of pin. There was a cold little click from inside the mechanism as it armed itself.

Jon could do nothing except watch, even his vocal diaphragm was locked by the magnetic field. He had more than a suspicion however that he was involved in something other than a "secret business deal." He cursed his own stupidity for walking blindly into the situation.

The magnetic field cut off and he instantly raced his extensor motors to leap forward. Coleman took a plastic box out of his pocket and held his thumb over a switch inset into its top.

"Don't make any quick moves, junk-yard, this little transmitter is keyed to a receiver in that bomb on your leg. One touch of my thumb, up you go in a cloud of smoke and come down in a shower of nuts and bolts." He signalled to Druce who opened a closet door. "And in case you want to be heroic, just think of him."

Coleman jerked his thumb at the sodden shape on the floor; a filthily attired man of indistinguishable age whose only interesting feature was the black bomb strapped tightly across his chest. He peered unseeingly from red-rimmed eyes and raised the almost empty whiskey bottle to his mouth. Coleman kicked the door shut.

"He's just some Bowery bum we dragged in, Venex, but that doesn't make any difference to you, does it? He's human—and a robot can't kill *anybody*! That rummy has a bomb on him tuned to the same frequency as yours, if you don't play ball with us he gets a two-foot hole blown in his chest."

Coleman was right, Jon didn't dare make any false moves. All of his early mental training as well as Circuit 92 sealed inside his brain case would prevent him from harming a human being. He felt trapped, caught by these people for some unknown purpose.

Coleman had pushed back a tarpaulin to disclose a ragged hole in the concrete floor, the opening extended into the earth below. He waved Jon over.

"The tunnel is in good shape for about thirty feet, then you'll find a fall. Clean all the rock and dirt out until you break through into the storm sewer, then come back. And you better be alone. If you tip the cops both you and the old stew go out together—now move."

The shaft had been dug recently and shored with packing crates from the warehouse overhead. It ended abruptly in a wall of fresh sand and stone. Jon began shoveling it into the little wheelbarrow they had given him.

He had emptied four barrow loads and was filling the fifth when he uncovered the hand, a robot's hand made of green metal. He turned his headlight power up and examined the hand closely, there could be no doubt about it. These gaskets on the joints, the rivet pattern at the base of the thumb meant only one thing, it was the dismembered hand of a Venex robot.

Quickly, yet gently, he shoveled away the rubble behind the hand and unearthed the rest of the robot. The torso was crushed and the power circuits shorted, battery acid was dripping from an ugly rent in the side. With infinite care Jon snapped the few remaining wires that joined the neck to the body and laid the green head on the barrow. It stared at him like a skull, the shutters completely dilated, but no glow of life from the tubes behind them.

He was scraping the mud from the number on the battered chestplate when Druce lowered himself into the tunnel and flashed the brilliant beam of a hand-spot down its length.

"Stop playing with that junk and get digging—or you'll end up the same as him. This tunnel has gotta be through by tonight."

Jon put the dismembered parts on the barrow with the sand and rock and pushed the whole load back up the tunnel, his thoughts running in unhappy circles. A dead robot was a terrible thing, and one of his family too. But there was something wrong about this robot, something that was quite inexplicable, the number on the plate had been "17," yet he remembered only too well the day that a water-shorted motor had killed Venex 17 in the Orange Sea.

It took Jon four hours to drive the tunnel as far as the ancient granite wall of the storm sewer. Druce gave him a short pinch bar and he levered out enough of the big blocks to make a hole large enough to let him through into the sewer.

When he climbed back into the office he tried to look casual as he dropped the pinch bar to the floor by his feet and seated himself on the pile of rubble in the corner. He moved around to make a comfortable seat for himself and his fingers grabbed the severed neck of Venex 17.

Coleman swiveled around in his chair and squinted at the wall clock. He checked the time against his tie-pin watch, with a grunt of satisfaction he turned back and stabbed a finger at Jon.

"Listen, you green junk-pile, at 1900 hours you're going to do a job, and there aren't going to be any slip ups. You go down that sewer and into the Hudson River. The outlet is under water, so you won't be seen from the docks. Climb down to the bottom and walk 200 yards north, that should put you just under a ship. Keep your eyes open, *but don't show any lights*! About halfway down the keel of the ship you'll find a chain hanging.

"Climb the chain, pull loose the box that's fastened to the hull at the top and bring it back here. No mistakes—or you know what happens."

Jon nodded his head. His busy fingers had been separating the wires in the amputated neck. When they had been straightened and put into a row he memorized their order with one flashing glance.

He ran over the color code in his mind and compared it with the memorized leads. The twelfth wire was the main cranial power lead, number six was the return wire.

With his precise touch he separated these two from the pack and glanced idly around the room. Druce was dozing on a chair in the opposite corner. Coleman was talking on the phone, his voice occasionally rising in a petulant whine. This wasn't interfering with his attention to Jon—and the radio switch still held tightly in left hand.

Jon's body blocked Coleman's vision, as long as Druce stayed asleep he would be able to work on the head unobserved. He activated a relay in his forearm and there was a click as the waterproof cover on an exterior socket swung open. This was a power outlet from his battery that was used to operate motorized tools and lights underwater.

If Venex 17's head had been severed for less than three weeks he could reactivate it. Every robot had a small storage battery inside his skull, if the power to the brain was cut off the battery would provide the minimum standby current to keep the brain alive. The robe would be unconscious until full power was restored.

Jon plugged the wires into his arm-outlet and slowly raised the current to operating level. There was a tense moment of waiting, then 17's eye shutters suddenly closed. When they opened again the eye tubes were glowing warmly. They swept the room with one glance then focused on Jon.

The right shutter clicked shut while the other began opening and closing in rapid fashion. It was International code—being sent as fast as the solenoid could be operated. Jon concentrated on the message.

Telephone—call emergency operator—tell her "signal 14" help will—

The shutter stopped in the middle of a code group, the light of reason dying from the eyes.

For one instant Jon's heart leaped in panic, until he realized that 17 had deliberately cut the power. Druce's harsh voice rasped in his ear.

"What you doing with that? None of your funny robot tricks. I know your kind, plotting all kinds of things in them tin domes." His voice trailed off into a stream of incomprehensible profanity. With sudden spite he lashed his foot out and sent 17's head crashing against the wall.

The dented, green head rolled to a stop at Jon's feet, the face staring up at him in mute agony. It was only Circuit 92 that prevented him from injuring a *human.* As his motors revved up to send him hurtling forward the control relays clicked open. He sank against the debris, paralyzed for the instant. As soon as the rush of anger was gone he would regain control of his body.

They stood as if frozen in a tableau. The robot slumped backward, the man leaning forward, his face twisted with unreasoning hatred. The head lay between them like a symbol of death.

Coleman's voice cut through the air of tenseness like a knife.

"*Druce*, stop playing with the grease-can and get down to the main door to let Little Willy and his junk-brokers in. You can have it all to yourself afterward."

The angry man turned reluctantly, but pushed out of the door at Coleman's annoyed growl. Jon sat down against the wall, his mind sorting out the few facts with lightning precision. There was no room in his thoughts for Druce, the man had become just one more factor in a complex problem.

Call the emergency operator—that meant this was no local matter, responsible authorities must be involved. Only the government could be behind a thing as major as this. Signal 14—that inferred a complex set of arrangements, forces that could swing into action at a moment's notice. There was no indication where this might lead, but the only thing to do was to get out of here and make that phone call. And quick. Druce was bringing in more people, junk-brokers, whatever they were. Any action that he took would have to be done before they returned.

Even as Jon followed this train of logic his fingers were busy. Palming a wrench, he was swiftly loosening the main retaining nut on his hip joint. It dropped free in his hand, only the pivot pin remained now to hold his leg on. He climbed slowly to his feet and moved towards Coleman's desk.

"Mr. Coleman, sir, it's time to go down to the ship now, should I leave now, sir?"

Jon spoke the words slowly as he walked forward, apparently going to the door, but angling at the same time towards the plump man's desk.

"You got thirty minutes yet, go sit—*say...*!"

The words were cut off. Fast as a human reflex is, it is the barest crawl compared to the lightning action of electronic reflex. At the instant Coleman was first aware of Jon's motion, the robot had finished his leap and lay sprawled across the desk, his leg off at the hip and clutched in his hand.

"YOU'LL KILL YOURSELF IF YOU TOUCH THE BUTTON!"

The words were part of the calculated plan. Jon bellowed them in the startled man's ear as he stuffed the dismembered leg down the front of the man's baggy slacks. It had the desired effect, Coleman's finger stabbed at the button but stopped before it made contact. He stared down with bulging eyes at the little black box of death peeping out of his waistband.

Jon hadn't waited for the reaction. He pushed backward from the desk and stopped to grab the stolen pinch bar off the floor. A mighty one-legged leap brought him to the locked closet; he stabbed the bar into the space between the door and frame and heaved.

Coleman was just starting to struggle the bomb out of his pants when the action was over. The closet open, Jon seized the heavy strap holding the second bomb on the rummy's chest and snapped it like a thread. He threw the bomb into Coleman's corner, giving the man one more thing to worry about. It had cost him a leg, but Jon had escaped the bomb threat without injuring a human. Now he had to get to a phone and make that call.

Coleman stopped tugging at the bomb and plunged his hand into the desk drawer for a gun. The returning men would block the door soon, the only other exit from the room was a frosted-glass window that opened onto the mammoth bay of the warehouse.

Jon Venex plunged through the window in a welter of flying glass. The heavy thud of a recoilless .75 came from the room behind him and a foot-long section of metal window frame leaped outward. Another slug screamed by the robot's head as he scrambled toward the rear door of the warehouse.

He was a bare thirty feet away from the back entrance when the giant door hissed shut on silent rollers. All the doors would have closed at the same time, the thud of running feet indicated that they would be guarded as well. Jon hopped a section of packing cases and crouched out of sight.

He looked up over his head, there stretched a webbing of steel supports, crossing and recrossing until they joined the flat expanse of the roof. To human eyes the shadows there deepened into obscurity, but the infra-red from a network of steam pipes gave Jon all the illumination he needed.

The men would be quartering the floor of the warehouse soon, his only chance to escape recapture or death would be over their heads. Besides this, he was hampered by the loss of his leg. In the rafters he could use his arms for faster and easier travel.

Jon was just pulling himself up to one of the topmost cross beams when a hoarse shout from below was followed by a stream of bullets. They tore through the thin roof, one slug clanged off the steel beam under his body. Waiting until three of the newcomers had started up a nearby ladder, Jon began to quietly work his way towards the back of the building.

Safe for the moment, he took stock of his position. The men were spread out through the building, it could only be a matter of time before they found him. The doors were all locked and—he had made a complete circuit of the building to be sure—there were no windows that he could force—the windows were bolted as well. If he could call the emergency operator the unknown friends of Venex 17 might come to his aid. This, however, was out of the question. The only phone in the building was on Coleman's desk. He had traced the leads to make sure.

His eyes went automatically to the cables above his head. Plastic gaskets were set in the wall of the building, through them came the power and phone lines. The phone line! That was all he needed to make a call.

With smooth, fast motions he reached up and scratched a section of wire bare. He laughed to himself as he slipped the little microphone out of his left ear. Now he was half deaf as well as half lame—he was literally giving himself to this cause. He would have to remember the pun to tell Alec Diger later, if there was a later. Alec had a profound weakness for puns.

Jon attached jumpers to the mike and connected them to the bare wire. A touch of the ammeter showed that no one was on the line. He waited a few moments to be sure he had a dial tone then sent the eleven carefully spaced pulses that would connect him with the local operator. He placed the mike close to his mouth.

"Hello, operator. Hello, operator. I cannot hear you so do not answer. Call the emergency operator—signal 14, I repeat—signal 14."

Jon kept repeating the message until the searching men began to approach his position. He left the mike connected—the men wouldn't notice it in the dark but the open line would give the unknown powers his exact location. Using his fingertips he did a careful traverse on an I-beam to an alcove in the farthest corner of the room. Escape was impossible, all he could do was stall for time.

"Mr. Coleman, I'm sorry I ran away." With the volume on full his voice rolled like thunder from the echoing walls.

He could see the men below twisting their heads vainly to find the source.

"If you let me come back and don't kill me I will do your work. I was afraid of the bomb, but now I am afraid of the guns." It sounded a little infantile, but he was pretty sure none of those present had any sound knowledge of robotic intelligence.

"Please let me come back ... sir!" He had almost forgotten the last word, so he added another "Please, sir!" to make up.

Coleman needed that package under the boat very badly, he would promise anything to get it. Jon had no doubts as to his eventual fate, all he could hope to do was kill time in the hopes that the phone message would bring aid.

"Come on down, Junky, I won't be mad at you—if you follow directions." Jon could hear the hidden anger in his voice, the unspoken hatred for a robe who dared lay hands on him.

The descent wasn't difficult, but Jon did it slowly with much apparent discomfort. He hopped into the center of the floor—leaning on the cases as if for support. Coleman and Druce were both there as well as a group of hard-eyed newcomers. They raised their guns at his approach but Coleman stopped them with a gesture.

"This is *my* robe, boys, I'll see to it that he's happy."

He raised his gun and shot Jon's remaining leg off. Twisted around by the blast, Jon fell helplessly to the floor. He looked up into the smoking mouth of the .75.

"Very smart for a tin-can, but not smart enough. We'll get the junk on the boat some other way, some way that won't mean having you around under foot." Death looked out of his narrowed eyes.

Less than two minutes had passed since Jon's call. The watchers must have been keeping 24 hour stations waiting for Venex 17's phone message.

The main door went down with the sudden scream of torn steel. A whippet tank crunched over the wreck and covered the group with its multiple pom-poms. They were an instant too late, Coleman pulled the trigger.

Jon saw the tensing trigger finger and pushed hard against the floor. His head rolled clear but the bullet tore through his shoulder. Coleman didn't have a chance for a second shot, there was a fizzling hiss from the tank and the riot ports released a flood of tear gas. The stricken men never saw the gas-masked police that poured in from the street.

Jon lay on the floor of the police station while a tech made temporary repairs on his leg and shoulder. Across the room Venex 17 was moving his new body with evident pleasure.

"Now this really feels like *something*! I was sure my time was up when that land slip caught me. But maybe I ought to start from the beginning." He stamped across the room and shook Jon's inoperable hand.

"The name is Wil Counter-4951L3, not that *that* means much any more. I've worn so many different bodies that I forget what I originally looked like. I went right from factory-school to a police training school—and I have been on the job ever since—Force of Detectives, Sergeant Jr. grade, Investigation Department. I spend most of my time selling candy bars or newspapers, or serving drinks in crumb joints. Gather information, make reports and keep tab on guys for other departments.

"This last job—and I'm sorry I had to use a Venex identity, I don't think I brought any dishonor to your family—I was on loan to the Customs department. Seems a ring was bringing uncut junk—heroin—into the country. F.B.I. tabbed all the operators here, but no one knew how the stuff got in. When Coleman, he's the local big-shot, called the agencies for an underwater robot, I was packed into a new body and sent running.

"I alerted the squad as soon as I started the tunnel, but the damned thing caved in on me before I found out what ship was doing the carrying. From there on you know what happened.

"Not knowing I was out of the game the squad sat tight and waited. The hop merchants saw a half million in snow sailing back to the old country so they had you dragged in as a replacement. You made the phone call and the cavalry rushed in at the last moment to save two robots from a rusty grave."

Jon, who had been trying vainly to get in a word, saw his chance as Wil Counter turned to admire the reflection of his new figure in a window.

"You shouldn't be telling me those things—about your police investigations and department operations. Isn't this information supposed to be secret? Specially from robots!"

"Of course it is!" was Wil's airy answer. "Captain Edgecombe—he's the head of my department—is an expert on all kinds of blackmail. I'm supposed to tell you so much confidential police business that you'll have to either join the department or be shot as a possible informer." His laughter wasn't shared by the bewildered Jon.

"Truthfully, Jon, we need you and can use you. Robes that can think fast and act fast aren't easy to find. After hearing about the tricks you pulled in that warehouse, the Captain swore to decapitate me permanently if I couldn't get you to join up. Do you need a job? Long hours, short pay—but guaranteed to never get boring."

Wil's voice was suddenly serious. "You saved my life, Jon—those snowbirds would have left me in that sandpile until all hell froze over. I'd like you for a mate, I think we could get along well together." The gay note came back into his voice, "And besides that, I may be able to save your life some day—I hate owing debts."

The tech was finished, he snapped his tool box shut and left. Jon's shoulder motor was repaired now, he sat up. When they shook hands this time it was a firm clasp. The kind you know will last awhile.

Jon stayed in an empty cell that night. It was gigantic compared to the hotel and barrack rooms he was used to.

He wished that he had his missing legs so he could take a little walk up and down the cell. He would have to wait until the morning. They were going to fix him up then before he started the new job.

He had recorded his testimony earlier and the impossible events of the past day kept whirling around in his head. He would think about it some other time, right now all he wanted to do was let his overworked circuits cool down, if he only had something to read, to focus his attention on. Then, with a start, he remembered the booklet. Everything had moved so fast that the earlier incident with the truck driver had slipped his mind completely.

He carefully worked it out from behind the generator shielding and opened the first page of *Robot Slaves in a World Economy*. A card slipped from between the pages and he read the short message on it.

PLEASE DESTROY THIS CARD AFTER READING

If you think there is truth in this book and would like to hear more, come to Room B, 107 George St. any Tuesday at 5 P.M.

The card flared briefly and was gone. But he knew that it wasn't only a perfect memory that would make him remember that message.

Survival Tactics

by Al Sevcik

Illustrated by Irving Novick

This story is taken from *Amazing Science Fiction Stories,* October 1958.

The robots were built to serve Man; to do his work, see to his comforts, make smooth his way. Then the robots figured out an additional service—putting Man out of his misery.

There was a sudden crash that hung sharply in the air, as if a tree had been hit by lightning some distance away. Then another. Alan stopped, puzzled. Two more blasts, quickly together, and the sound of a scream faintly.

Frowning, worrying about the sounds, Alan momentarily forgot to watch his step until his foot suddenly plunged into an ant hill, throwing him to the jungle floor. "Damn!" He cursed again, for the tenth time, and stood uncertainly in the dimness. From tall, moss-shrouded trees, wrist-thick vines hung quietly, scraping the spongy ground like the tentacles of some monstrous tree-bound octopus. Fitful little plants grew straggly in the shadows of the mossy trunks, forming a dense underbrush that made walking difficult. At midday some few of the blue sun's rays filtered through to the jungle floor, but now, late afternoon on the planet, the shadows were long and gloomy.

Alan peered around him at the vine-draped shadows, listening to the soft rustlings and faint twig-snappings of life in the jungle. Two short, popping sounds echoed across the stillness, drowned out almost immediately and silenced by an explosive crash. Alan started, "Blaster fighting! But it can't be!"

Suddenly anxious, he slashed a hurried X in one of the trees to mark his position then turned to follow a line of similar marks back through the jungle. He tried to run, but vines blocked his way and woody shrubs caught at his legs, tripping him and holding him back. Then, through the trees he saw the clearing of the camp site, the temporary home for the scout ship and the eleven men who, with Alan, were the only humans on the jungle planet, Waiamea.

Stepping through the low shrubbery at the edge of the site, he looked across the open area to the two temporary structures, the camp headquarters where the power supplies and the computer were; and the sleeping quarters. Beyond, nose high, stood the silver scout ship that had brought the advance exploratory party of scientists and technicians to Waiamea three days before. Except for a few of the killer robots rolling slowly around the camp site on their quiet treads, there was no one about.

"So, they've finally got those things working." Alan smiled slightly. "Guess that means I owe Pete a bourbon-and-soda for sure. Anybody who can build a robot that hunts by homing in on animals' mind impulses ..." He stepped forward just as a roar of blue flame dissolved the branches of a tree, barely above his head.

Without pausing to think, Alan leaped back, and fell sprawling over a bush just as one of the robots rolled silently up from the right, lowering its blaster barrel to aim directly at his head. Alan froze. "My God, Pete built those things wrong!"

Suddenly a screeching whirlwind of claws and teeth hurled itself from the smoldering branches and crashed against the robot, clawing insanely at the antenna and blaster barrel. With an awkward jerk the robot swung around and fired its blaster, completely dissolving the lower half of the cat creature which had clung across the barrel. But the back pressure of the cat's body overloaded the discharge circuits. The robot started to shake, then clicked sharply as an overload relay snapped and shorted the blaster cells. The killer turned and rolled back towards the camp, leaving Alan alone.

Shakily, Alan crawled a few feet back into the undergrowth where he could lie and watch the camp, but not himself be seen. Though visibility didn't make any difference to the robots, he felt safer, somehow, hidden. He knew now what the shooting sounds had been and why there hadn't been anyone around the camp site. A charred blob lying in the grass of the clearing confirmed his hypothesis. His stomach felt sick.

"I suppose," he muttered to himself, "that Pete assembled these robots in a batch and then activated them all at once, probably never living to realize that they're tuned to pick up human brain waves, too. Damn! Damn!" His eyes blurred and he slammed his fist into the soft earth.

When he raised his eyes again the jungle was perceptibly darker. Stealthy rustlings in the shadows grew louder with the setting sun. Branches snapped unaccountably in the trees overhead and every now and then leaves or a twig fell softly to the ground, close to where he lay. Reaching into his jacket, Alan fingered his pocket blaster. He pulled it out and held it in his right hand. "This pop gun wouldn't even singe a robot, but it just might stop one of those pumas."

They said the blast with your name on it would find you anywhere. This looked like Alan's blast.

Slowly Alan looked around, sizing up his situation. Behind him the dark jungle rustled forbiddingly. He shuddered. "Not a very healthy spot to spend the night. On the other hand, I certainly can't get to the camp with a pack of mind-activated mechanical killers running around. If I can just hold out until morning, when the big ship arrives ... The big ship! Good Lord, Peggy!" He turned white; oily sweat punctuated his forehead. Peggy, arriving tomorrow with the other colonists, the wives and kids! The metal killers, tuned to blast any living flesh, would murder them the instant they stepped from the ship!

A pretty girl, Peggy, the girl he'd married just three weeks ago. He still couldn't believe it. It was crazy, he supposed, to marry a girl and then take off for an unknown planet, with her to follow, to try to create a home in a jungle clearing. Crazy maybe, but Peggy and her green eyes that changed color with the light, with her soft brown hair, and her happy smile, had ended thirty years of loneliness and had, at last, given him a reason for living. "Not to be killed!" Alan unclenched his fists and wiped his palms, bloody where his fingernails had dug into the flesh.

There was a slight creak above him like the protesting of a branch too heavily laden. Blaster ready, Alan rolled over onto his back. In the movement, his elbow struck the top of a small earthy mound and he was instantly engulfed in a swarm of locust-like insects that beat disgustingly against his eyes and mouth. "Fagh!" Waving his arms before his face he jumped up and backwards, away from the bugs. As he did so, a dark shapeless thing plopped from the trees onto the spot where he had been lying stretched out. Then, like an ambient fungus, it slithered off into the jungle undergrowth.

For a split second the jungle stood frozen in a brilliant blue flash, followed by the sharp report of a blaster. Then another. Alan whirled, startled. The planet's double moon had risen and he could see a robot rolling slowly across the clearing in his general direction, blasting indiscriminately at whatever mind impulses came within its pickup range, birds, insects, anything. Six or seven others also left the camp headquarters area and headed for the jungle, each to a slightly different spot.

Apparently the robot hadn't sensed him yet, but Alan didn't know what the effective range of its pickup devices was. He began to slide back into the jungle. Minutes later, looking back he saw that the machine, though several hundred yards away, had altered its course and was now headed directly for him.

His stomach tightened. Panic. The dank, musty smell of the jungle seemed for an instant to thicken and choke in his throat. Then he thought of the big ship landing in the morning, settling down slowly after a lonely two-week voyage. He thought of a brown-haired girl crowding with the others to the gangway, eager to embrace the new planet, and the next instant a charred nothing, unrecognizable, the victim of a design error or a misplaced wire in a machine. "I have to try," he said aloud. "I have to try." He moved into the blackness.

Powerful as a small tank, the killer robot was equipped to crush, slash, and burn its way through undergrowth. Nevertheless, it was slowed by the larger trees and the thick, clinging vines, and Alan found that he could manage to keep ahead of it, barely out of blaster range. Only, the robot didn't get tired. Alan did.

The twin moons cast pale, deceptive shadows that wavered and danced across the jungle floor, hiding debris that tripped him and often sent him sprawling into the dark. Sharp-edged growths tore at his face and clothes, and insects attracted by the blood matted against his pants and shirt. Behind, the robot crashed imperturbably after him, lighting the night with fitful blaster flashes as some winged or legged life came within its range.

There was movement also, in the darkness beside him, scrapings and rustlings and an occasional low, throaty sound like an angry cat. Alan's fingers tensed on his pocket blaster. Swift shadowy forms moved quickly in the shrubs and the growling became suddenly louder. He fired twice, blindly, into the undergrowth. Sharp screams punctuated the electric blue discharge as a pack of small feline creatures leaped snarling and clawing back into the night.

Mentally, Alan tried to figure the charge remaining in his blaster. There wouldn't be much. "Enough for a few more shots, maybe. Why the devil didn't I load in fresh cells this morning!"

The robot crashed on, louder now, gaining on the tired human. Legs aching and bruised, stinging from insect bites, Alan tried to force himself to run holding his hands in front of him like a child in the dark. His foot tripped on a barely visible insect hill and a winged swarm exploded around him. Startled, Alan jerked sideways, crashing his head against a tree. He clutched at the bark for a second, dazed, then his knees buckled. His blaster fell into the shadows.

The robot crashed loudly behind him now. Without stopping to think, Alan fumbled along the ground after his gun, straining his eyes in the darkness. He found it just a couple of feet to one side, against the base of a small bush. Just as his fingers closed upon the barrel his other hand slipped into something sticky that splashed over his forearm. He screamed in pain and leaped back, trying frantically to wipe the clinging, burning blackness off his arm. Patches of black scraped off onto branches and vines, but the rest spread slowly over his arm as agonizing as hot acid, or as flesh being ripped away layer by layer.

Almost blinded by pain, whimpering, Alan stumbled forward. Sharp muscle spasms shot from his shoulder across his back and chest. Tears streamed across his cheeks.

A blue arc slashed at the trees a mere hundred yards behind. He screamed at the blast. "Damn you, Pete! Damn your robots! Damn, damn ... Oh, Peggy!" He stepped into emptiness.

Coolness. Wet. Slowly, washed by the water, the pain began to fall away. He wanted to lie there forever in the dark, cool, wetness. For ever, and ever, and ... The air thundered.

In the dim light he could see the banks of the stream, higher than a man, muddy and loose. Growing right to the edge of the banks, the jungle reached out with hairy, disjointed arms as if to snag even the dirty little stream that passed so timidly through its domain.

Alan, lying in the mud of the stream bed, felt the earth shake as the heavy little robot rolled slowly and inexorably towards him. "The Lord High Executioner," he thought, "in battle dress." He tried to stand but his legs were almost too weak and his arm felt numb. "I'll drown him," he said aloud. "I'll drown the Lord High Executioner." He laughed. Then his mind cleared. He remembered where he was.

Alan trembled. For the first time in his life he understood what it was to live, because for the first time he realized that he would sometime die. In other times and circumstances he might put it off for a while, for months or years, but eventually, as now, he would have to watch, still and helpless, while death came creeping. Then, at thirty, Alan became a man.

"Dammit, no law says I have to flame-out *now*!" He forced himself to rise, forced his legs to stand, struggling painfully in the shin-deep ooze. He worked his way to the bank and began to dig frenziedly, chest high, about two feet below the edge.

His arm where the black thing had been was swollen and tender, but he forced his hands to dig, dig, dig, cursing and crying to hide the pain, and biting his lips, ignoring the salty taste of blood. The soft earth crumbled under his hands until he had a small cave about three feet deep in the bank. Beyond that the soil was held too tightly by the roots from above and he had to stop.

The air crackled blue and a tree crashed heavily past Alan into the stream. Above him on the bank, silhouetting against the moons, the killer robot stopped and its blaster swivelled slowly down. Frantically, Alan hugged the bank as a shaft of pure electricity arced over him, sliced into the water, and exploded in a cloud of steam. The robot shook for a second, its blaster muzzle lifted erratically and for an instant it seemed almost out of control, then it quieted and the muzzle again pointed down.

Pressing with all his might, Alan slid slowly along the bank inches at a time, away from the machine above. Its muzzle turned to follow him but the edge of the bank blocked its aim. Grinding forward a couple of feet, slightly overhanging the bank, the robot fired again. For a split second Alan seemed engulfed in flame; the heat of hell singed his head and back, and mud boiled in the bank by his arm.

Again the robot trembled. It jerked forward a foot and its blaster swung slightly away. But only for a moment. Then the gun swung back again.

Suddenly, as if sensing something wrong, its tracks slammed into reverse. It stood poised for a second, its treads spinning crazily as the earth collapsed underneath it, where Alan had dug, then it fell with a heavy splash into the mud, ten feet from where Alan stood.

Without hesitation Alan threw himself across the blaster housing, frantically locking his arms around the barrel as the robot's treads churned furiously in the sticky mud, causing it to buck and plunge like a Brahma bull. The treads stopped and the blaster jerked upwards wrenching Alan's arms, then slammed down. Then the whole housing whirled around and around, tilting alternately up and down like a steel-skinned water monster trying to dislodge a tenacious crab, while Alan, arms and legs wrapped tightly around the blaster barrel and housing, pressed fiercely against the robot's metal skin.

Slowly, trying to anticipate and shift his weight with the spinning plunges, Alan worked his hand down to his right hip. He fumbled for the sheath clipped to his belt, found it, and extracted a stubby hunting knife. Sweat and blood in his eyes, hardly able to move on the wildly swinging turret, he felt down the sides to the thin crack between the revolving housing and the stationary portion of the robot. With a quick prayer he jammed in the knife blade—and was whipped headlong into the mud as the turret literally snapped to a stop.

The earth, jungle and moons spun in a pinwheeled blur, slowed, and settled to their proper places. Standing in the sticky, sweet-smelling ooze, Alan eyed the robot apprehensively. Half buried in mud, it stood quiet in the shadowy light except for an occasional, almost spasmodic jerk of its blaster barrel. For the first time that night Alan allowed himself a slight smile. "A blade in the old gear box, eh? How does that feel, boy?"

He turned. "Well, I'd better get out of here before the knife slips or the monster cooks up some more tricks with whatever it's got for a brain." Digging little footholds in the soft bank, he climbed up and stood once again in the rustling jungle darkness.

"I wonder," he thought, "how Pete could cram enough brain into one of those things to make it hunt and track so perfectly." He tried to visualize the computing circuits needed for the operation of its tracking mechanism alone. "There just isn't room for the electronics. You'd need a computer as big as the one at camp headquarters."

In the distance the sky blazed as a blaster roared in the jungle. Then Alan heard the approaching robot, crunching and snapping its way through the undergrowth like an onrushing forest fire. He froze. "Good Lord! They communicate with each other! The one I jammed must be calling others to help."

He began to move along the bank, away from the crashing sounds. Suddenly he stopped, his eyes widened. "Of course! Radio! I'll bet anything they're automatically controlled by the camp computer. That's where their brain is!" He paused. "Then, if that were put out of commission ..." He jerked away from the bank and half ran, half pulled himself through the undergrowth towards the camp.

Trees exploded to his left as another robot fired in his direction, too far away to be effective but churning towards him through the blackness.

Alan changed direction slightly to follow a line between the two robots coming up from either side, behind him. His eyes were well accustomed to the dark now, and he managed to dodge most of the shadowy vines and branches before they could snag or trip him. Even so, he stumbled in the wiry underbrush and his legs were a mass of stinging slashes from ankle to thigh.

The crashing rumble of the killer robots shook the night behind him, nearer sometimes, then falling slightly back, but following constantly, more unshakable than bloodhounds because a man can sometimes cover a scent, but no man can stop his thoughts. Intermittently, like photographers' strobes, blue flashes would light the jungle about him. Then, for seconds afterwards his eyes would see dancing streaks of yellow and sharp multi-colored pinwheels that alternately shrunk and expanded as if in a surrealist's nightmare. Alan would have to pause and squeeze his eyelids tight shut before he could see again, and the robots would move a little closer.

To his right the trees silhouetted briefly against brilliance as a third robot slowly moved up in the distance. Without thinking, Alan turned slightly to the left, then froze in momentary panic. "I should be at the camp now. Damn, what direction am I going?" He tried to think back, to visualize the twists and turns he'd taken in the jungle. "All I need is to get lost."

He pictured the camp computer with no one to stop it, automatically sending its robots in wider and wider forays, slowly wiping every trace of life from the planet. Technologically advanced machines doing the job for which they were built, completely, thoroughly, without feeling, and without human masters to separate sense from futility. Finally parts would wear out, circuits would short, and one by one the killers would crunch to a halt. A few birds would still fly then, but a unique animal life, rare in the universe, would exist no more. And the bones of children, eager girls, and their men would also lie, beside a rusty hulk, beneath the alien sun.

"Peggy!"

As if in answer, a tree beside him breathed fire, then exploded. In the brief flash of the blaster shot, Alan saw the steel glint of a robot only a hundred yards away, much nearer than he had thought. "Thank heaven for trees!" He stepped back, felt his foot catch in something, clutched futilely at some leaves and fell heavily.

Pain danced up his leg as he grabbed his ankle. Quickly he felt the throbbing flesh. "Damn the rotten luck, anyway!" He blinked the pain tears from his eyes and looked up—into a robot's blaster, jutting out of the foliage, thirty yards away.

Instinctively, in one motion Alan grabbed his pocket blaster and fired. To his amazement the robot jerked back, its gun wobbled and started to tilt away. Then, getting itself under control, it swung back again to face Alan. He fired again, and again the robot reacted. It seemed familiar somehow. Then he remembered the robot on the river bank, jiggling and swaying for seconds after each shot. "Of course!" He cursed himself for missing the obvious. "The blaster static blanks out radio transmission from the computer for a few seconds. They even do it to themselves!"

Firing intermittently, he pulled himself upright and hobbled ahead through the bush. The robot shook spasmodically with each shot, its gun tilted upward at an awkward angle.

Then, unexpectedly, Alan saw stars, real stars brilliant in the night sky, and half dragging his swelling leg he stumbled out of the jungle into the camp clearing. Ahead, across fifty yards of grass stood the headquarters building, housing the robot-controlling computer. Still firing at short intervals he started across the clearing, gritting his teeth at every step.

Straining every muscle in spite of the agonizing pain, Alan forced himself to a limping run across the uneven ground, carefully avoiding the insect hills that jutted up through the grass. From the corner of his eye he saw another of the robots standing shakily in the dark edge of the jungle waiting, it seemed, for his small blaster to run dry.

"Be damned! You can't win now!" Alan yelled between blaster shots, almost irrational from the pain that ripped jaggedly through his leg. Then it happened. A few feet from the building's door his blaster quit. A click. A faint hiss when he frantically jerked the trigger again and again, and the spent cells released themselves from the device, falling in the grass at his feet. He dropped the useless gun.

"No!" He threw himself on the ground as a new robot suddenly appeared around the edge of the building a few feet away, aimed, and fired. Air burned over Alan's back and ozone tingled in his nostrils.

Blinding itself for a few seconds with its own blaster static, the robot paused momentarily, jiggling in place. In this instant, Alan jammed his hands into an insect hill and hurled the pile of dirt and insects directly at the robot's antenna. In a flash, hundreds of the winged things erupted angrily from the hole in a swarming cloud, each part of which was a speck of life transmitting mental energy to the robot's pickup devices.

Confused by the sudden dispersion of mind impulses, the robot fired erratically as Alan crouched and raced painfully for the door. It fired again, closer, as he fumbled with the lock release. Jagged bits of plastic and stone ripped past him, torn loose by the blast.

Frantically, Alan slammed open the door as the robot, sensing him strongly now, aimed point blank. He saw nothing, his mind thought of nothing but the red-clad safety switch mounted beside the computer. Time stopped. There was nothing else in the world. He half-jumped, half-fell towards it, slowly, in tenths of seconds that seemed measured out in years.

The universe went black.

Later. Brilliance pressed upon his eyes. Then pain returned, a multi-hurting thing that crawled through his body and dragged ragged tentacles across his brain. He moaned.

A voice spoke hollowly in the distance. "He's waking. Call his wife."

Alan opened his eyes in a white room; a white light hung over his head. Beside him, looking down with a rueful smile, stood a young man wearing space medical insignia. "Yes," he acknowledged the question in Alan's eyes, "you hit the switch. That was three days ago. When you're up again we'd all like to thank you."

Suddenly a sobbing-laughing green-eyed girl was pressed tightly against him. Neither of them spoke. They couldn't. There was too much to say.

Beside Still Waters

by Robert Sheckley

Illustrated by Virgil Finlay

This story is taken from *Amazing Stories* Oct.-Nov. 1953.

When people talk about getting away from it all, they are usually thinking about our great open spaces out west. But to science fiction writers, that would be practically in the heart of Times Square. When a man of the future wants solitude he picks a slab of rock floating in space four light years east of Andromeda. Here is a gentle little story about a man who sought the solitude of such a location. And who did he take along for company? None other than Charles the Robot.

Mark Rogers was a prospector, and he went to the asteroid belt looking for radioactives and rare metals. He searched for years, never finding much, hopping from fragment to fragment. After a time he settled on a slab of rock half a mile thick.

Rogers had been born old, and he didn't age much past a point. His face was white with the pallor of space, and his hands shook a little. He called his slab of rock Martha, after no girl he had ever known.

He made a little strike, enough to equip Martha with an air pump and a shack, a few tons of dirt and some water tanks, and a robot. Then he settled back and watched the stars.

The robot he bought was a standard-model all-around worker, with built-in memory and a thirty-word vocabulary. Mark added to that, bit by bit. He was something of a tinkerer, and he enjoyed adapting his environment to himself.

At first, all the robot could say was "Yes, sir," and "No, sir." He could state simple problems: "The air pump is laboring, sir." "The corn is budding, sir." He could perform a satisfactory salutation: "Good morning, sir."

Mark changed that. He eliminated the "sirs" from the robot's vocabulary; equality was the rule on Mark's hunk of rock. Then he dubbed the robot Charles, after a father he had never known.

As the years passed, the air pump began to labor a little as it converted the oxygen in the planetoid's rock into a breathable atmosphere. The air seeped into space, and the pump worked a little harder, supplying more.

The crops continued to grow on the tamed black dirt of the planetoid. Looking up, Mark could see the sheer blackness of the river of space, the floating points of the stars. Around him, under him, overhead, masses of rock drifted, and sometimes the starlight glinted from their black sides. Occasionally, Mark caught a glimpse of Mars or Jupiter. Once he thought he saw Earth.

Mark began to tape new responses into Charles. He added simple responses to cue words. When he said, "How does it look?" Charles would answer, "Oh, pretty good, I guess."

At first the answers were what Mark had been answering himself, in the long dialogue held over the years. But, slowly, he began to build a new personality into Charles.

Mark had always been suspicious and scornful of women. But for some reason he didn't tape the same suspicion into Charles. Charles' outlook was quite different.

"What do you think of girls?" Mark would ask, sitting on a packing case outside the shack, after the chores were done.

"Oh, I don't know. You have to find the right one." The robot would reply dutifully, repeating what had been put on its tape.

"I never saw a good one yet," Mark would say.

"Well, that's not fair. Perhaps you didn't look long enough. There's a girl in the world for every man."

"You're a romantic!" Mark would say scornfully. The robot would pause—a built-in pause—and chuckle a carefully constructed chuckle.

"I dreamed of a girl named Martha once," Charles would say. "Maybe if I would have looked, I would have found her."

And then it would be bedtime. Or perhaps Mark would want more conversation. "What do you think of girls?" he would ask again, and the discussion would follow its same course.

Charles grew old. His limbs lost their flexibility, and some of his wiring started to corrode. Mark would spend hours keeping the robot in repair.

"You're getting rusty," he would cackle.

"You're not so young yourself," Charles would reply. He had an answer for almost everything. Nothing involved, but an answer.

It was always night on Martha, but Mark broke up his time into mornings, afternoons and evenings. Their life followed a simple routine. Breakfast, from vegetables and Mark's canned store. Then the robot would work in the fields, and the plants grew used to his touch. Mark would repair the pump, check the water supply, and straighten up the immaculate shack. Lunch, and the robot's chores were usually finished.

The two would sit on the packing case and watch the stars. They would talk until supper, and sometimes late into the endless night.

In time, Mark built more complicated conversations into Charles. He couldn't give the robot free choice, of course, but he managed a pretty close approximation of it. Slowly, Charles' personality emerged. But it was strikingly different from Mark's.

Where Mark was querulous, Charles was calm. Mark was sardonic, Charles was naive. Mark was a cynic, Charles was an idealist. Mark was often sad; Charles was forever content.

And in time, Mark forgot he had built the answers into Charles. He accepted the robot as a friend, of about his own age. A friend of long years' standing.

"The thing I don't understand," Mark would say, "is why a man like you wants to live here. I mean, it's all right for me. No one cares about me, and I never gave much of a damn about anyone. But why you?"

"Here I have a whole world," Charles would reply, "where on Earth I had to share with billions. I have the stars, bigger and brighter than on Earth. I have all space around me, close, like still waters. And I have you, Mark."

"Now, don't go getting sentimental on me—"

"I'm not. Friendship counts. Love was lost long ago, Mark. The love of a girl named Martha, whom neither of us ever met. And that's a pity. But friendship remains, and the eternal night."

"You're a bloody poet," Mark would say, half admiringly. "A poor poet."

Time passed unnoticed by the stars, and the air pump hissed and clanked and leaked. Mark was fixing it constantly, but the air of Martha became increasingly rare. Although Charles labored in the fields, the crops, deprived of sufficient air, died.

Mark was tired now, and barely able to crawl around, even without the grip of gravity. He stayed in his bunk most of the time. Charles fed him as best he could, moving on rusty, creaking limbs.

"What do you think of girls?"

"I never saw a good one yet."

"Well, that's not fair."

Mark was too tired to see the end coming, and Charles wasn't interested. But the end was on its way. The air pump threatened to give out momentarily. There hadn't been any food for days.

"But why you?" Gasping in the escaping air. Strangling.

"Here I have a whole world—"

"Don't get sentimental—"

"And the love of a girl named Martha."

From his bunk Mark saw the stars for the last time. Big, bigger than ever, endlessly floating in the still waters of space.

"The stars ..." Mark said.

"Yes?"

"The sun?"

"—shall shine as now."

"A bloody poet."

"A poor poet."

"And girls?"

"I dreamed of a girl named Martha once. Maybe if—"

"What do you think of girls? And stars? And Earth?" And it was bedtime, this time forever.

Charles stood beside the body of his friend. He felt for a pulse once, and allowed the withered hand to fall. He walked to a corner of the shack and turned off the tired air pump.

The tape that Mark had prepared had a few cracked inches left to run. "I hope he finds his Martha," the robot croaked, and then the tape broke.

His rusted limbs would not bend, and he stood frozen, staring back at the naked stars. Then he bowed his head.

"The Lord is my shepherd," Charles said. "I shall not want. He maketh me to lie down in green pastures; he leadeth me ..."

A Spaceship Named McGuire

by Gordon Randall Garrett

Illustrated by Douglas

This story is from Analog, July 1961.

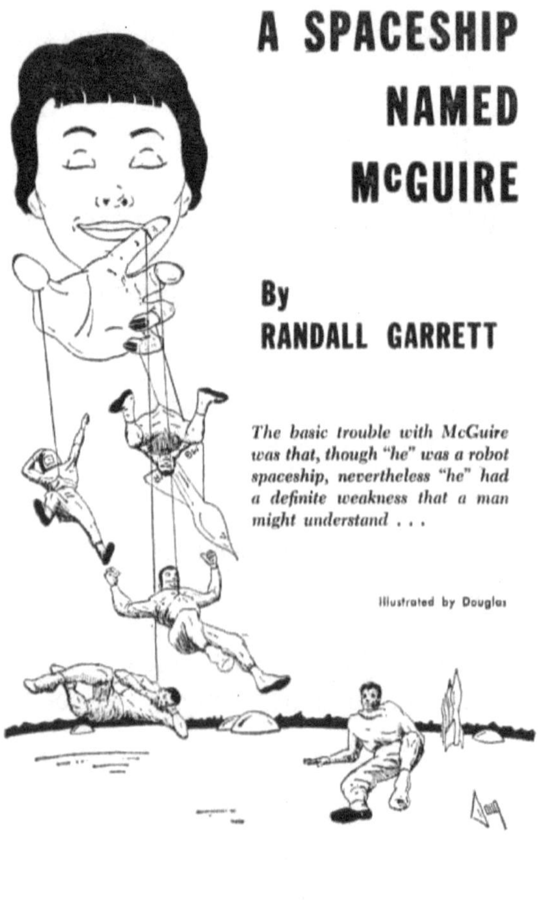

A SPACESHIP NAMED McGUIRE

By
RANDALL GARRETT

The basic trouble with McGuire was that, though "he" was a robot spaceship, nevertheless "he" had a definite weakness that a man might understand . . .

Illustrated by Douglas

The basic trouble with McGuire was that, though "he" was a robot spaceship, nevertheless "he" had a definite weakness that a man might understand....

No. Nobody ever deliberately named a spaceship that. The staid and stolid minds that run the companies which design and build spaceships rarely let their minds run to fancy. The only example I can think of is the unsung hero of the last century who had puckish imagination enough to name the first atomic-powered submarine *Nautilus*. Such minds are rare. Most minds equate dignity with dullness.

This ship happened to have a magnetogravitic drive, which automatically put it into the MG class. It also happened to be the first successful model to be equipped with a Yale robotic brain, so it was given the designation MG-YR-7—the first six had had more bugs in them than a Leopoldville tenement.

So somebody at Yale—another unsung hero—named the ship McGuire; it wasn't official, but it stuck.

The next step was to get someone to test-hop McGuire. They needed just the right man—quick-minded, tough, imaginative, and a whole slew of complementary adjectives. They wanted a perfect superman to test pilot their baby, even if they knew they'd eventually have to take second best.

It took the Yale Space Foundation a long time to pick the right man.

No, I'm not the guy who tested the McGuire.

I'm the guy who stole it.

Shalimar Ravenhurst is not the kind of bloke that very many people can bring themselves to like, and, in this respect, I'm like a great many people, if not more so. In the first place, a man has no right to go around toting a name like "Shalimar"; it makes names like "Beverly" and "Leslie" and "Evelyn" sound almost hairy chested. You want a dozen other reasons, you'll get them.

Shalimar Ravenhurst owned a little planetoid out in the Belt, a hunk of nickel-iron about the size of a smallish mountain with a gee-pull measurable in fractions of a centimeter per second squared. If you're susceptible to spacesickness, that kind of gravity is about as much help as aspirin would have been to Marie Antoinette. You get the feeling of a floor beneath you, but there's a distinct impression that it won't be there for long. It keeps trying to drop out from under you.

I dropped my flitterboat on the landing field and looked around without any hope of seeing anything. I didn't. The field was about the size of a football field, a bright, shiny expanse of rough-polished metal, carved and smoothed flat from the nickel-iron of the planetoid itself. It not only served as a landing field, but as a reflector beacon, a mirror that flashed out the sun's reflection as the planetoid turned slowly on its axis. I'd homed in on that beacon, and now I was sitting on it.

There wasn't a soul in sight. Off to one end of the rectangular field was a single dome, a hemisphere about twenty feet in diameter and half as high. Nothing else.

I sighed and flipped on the magnetic anchor, which grabbed hold of the metal beneath me and held the flitterboat tightly to the surface. Then I cut the drive, plugged in the telephone, and punched for "Local."

The automatic finder searched around for the Ravenhurst tickler signal, found it, and sent out a beep along the same channel.

I waited while the thing beeped twice. There was a click, and a voice said: "Raven's Rest. Yes?" It wasn't Ravenhurst.

I said: "This is Daniel Oak. I want to talk to Mr. Ravenhurst."

"Mr. Oak? But you weren't expected until tomorrow."

"Fine. I'm early. Let me talk to Ravenhurst."

"But Mr. Ravenhurst wasn't expecting you to—"

I got all-of-a-sudden exasperated. "Unless your instruments are running on secondhand flashlight batteries, you've known I was coming for the past half hour. I followed Ravenhurst's instructions not to use radio, but he should know I'm here by this time. He told me to come as fast as possible, and I followed those instructions, too. I always follow instructions when I'm paid enough.

"Now, I'm here; tell Ravenhurst I want to talk to him, or I'll simply flit back to Eros, and thank him much for a pretty retainer that didn't do him any good but gave me a nice profit for my trouble."

"One moment, please," said the voice.

It took about a minute and a half, which was about nine billion jiffies too long, as far as I was concerned.

Then another voice said: "Oak? Wasn't expecting you till tomorrow."

"So I hear. I thought you were in a hurry, but if you're not, you can just provide me with wine, women, and other necessities until tomorrow. That's above and beyond my fee, of course, since you're wasting my time, and I'm evidently not wasting yours."

I couldn't be sure whether the noise he made was a grunt or a muffled chuckle, and I didn't much care. "Sorry, Oak; I really didn't expect you so soon, but I do want to ... I want you to get started right away. Leave your flitterboat where it is; I'll have someone take care of it. Walk on over to the dome and come on in." And he cut off.

I growled something I was glad he didn't hear and hung up. I wished that I'd had a vision unit on the phone; I'd like to have seen his face. Although I knew I might not have learned much more from his expression than I had from his voice.

I got out of the flitterboat, and walked across the dome, my magnetic soles making subdued clicking noises inside the suit as they caught and released the metallic plain beneath me. Beyond the field, I was surrounded by a lumpy horizon and a black sky full of bright, hard stars.

The green light was on when I reached the door to the dome, so I opened it and went on in, closing it behind me. I flipped the toggle that began flooding the room with air. When it was up to pressure, a trap-door in the floor of the dome opened and a crew-cut, blond young man stuck his head up. "Mr. Oak?"

I toyed, for an instant, with the idea of giving him a sarcastic answer. Who else would it be? How many other visitors were running around on the surface of Raven's Rest?

Instead, I said: "That's right." My voice must have sounded pretty muffled to him through my fishbowl.

"Come on down, Mr. Oak. You can shuck your vac suit below."

I thought "below" was a pretty ambiguous term on a low-gee lump like this, but I followed him down the ladder. The ladder was a necessity for fast transportation; if I'd just tried to jump down from one floor to the next, it would've taken me until a month from next St. Swithin's Day to land.

The door overhead closed, and I could hear the pumps start cycling. The warning light turned red.

I took off my suit, hung it in a handy locker, showing that all I had on underneath was my skin-tight "union suit."

"All right if I wear this?" I asked the blond young man, "Or should I borrow a set of shorts and a jacket?" Most places in the Belt, a union suit is considered normal dress; a man never knows when he might have to climb into a vac suit—*fast*. But there are a few of the hoity-toity places on Eros and Ceres and a few of the other well-settled places where a man or woman is required to put on shorts and jacket before entering. And in good old New York City, a man and woman were locked up for "indecent exposure" a few months ago. The judge threw the case out of court, but he told them they were lucky they hadn't been picked up in Boston. It seems that the eye of the bluenose turns a jaundiced yellow at the sight of a union suit, and he sees red.

But there were evidently no bluenoses here. "Perfectly all right, Mr. Oak," the blond young man said affably. Then he coughed politely and added: "But I'm afraid I'll have to ask you to take off the gun."

I glanced at the holster under my armpit, walked back over to the locker, opened it, and took out my vac suit.

"Hey!" said the blond young man. "Where are you going?"

"Back to my boat," I said calmly. "I'm getting tired of this runaround already. I'm a professional man, not a hired flunky. If you'd called a doctor, you wouldn't tell him to leave his little black bag behind; if you'd called a lawyer, you wouldn't make him check his brief case. Or, if you did, he'd tell you to drop dead.

"I was asked to come here as fast as possible, and when I do, I'm told to wait till tomorrow. Now you want me to check my gun. The hell with you."

"Merely a safety precaution," said the blond young man worriedly.

"You think I'm going to shoot Ravenhurst, maybe? Don't be an idiot." I started climbing into my vac suit.

"Just a minute, please, Mr. Oak," said a voice from a hidden speaker. It was Ravenhurst, and he actually sounded apologetic. "You mustn't blame Mr. Feller; those are my standing orders, and I failed to tell Mr. Feller to make an exception in your case. The error was mine."

"I know," I said. "I wasn't blaming Mr. Feller. I wasn't even talking to him. I was addressing you."

"I believe you. Mr. Feller, our guest has gone to all the trouble of having a suit made with a space under the arm for that gun; I see no reason to make him remove it." A pause. "Again, Mr. Oak, I apologize. I really want you to take this job."

I was already taking off the vac suit again.

"But," Ravenhurst continued smoothly, "if I fail to live up to your ideas of courtesy again, I hope you'll forgive me in advance. I'm sometimes very forgetful, and I don't like it when a man threatens to leave my employ twice in the space of fifteen minutes."

"I'm not in your employ yet, Ravenhurst," I said. "If I accept the job, I won't threaten to quit again unless I mean to carry it through, and it would take a lot more than common discourtesy to make me do that. On the other hand, your brand of discourtesy is a shade above the common."

"I thank you for that, at least," said Ravenhurst. "Show him to my office, Mr. Feller."

The blond young man nodded wordlessly and led me from the room.

Walking under low-gee conditions is like nothing else in this universe. I don't mean trotting around on Luna; one-sixth gee is practically homelike in comparison. And zero gee is so devoid of orientation that it gives the sensation of falling endlessly until you get used to it. But a planetoid is in a different class altogether.

Remember that dream—almost everybody's had it—where you're suddenly able to fly? It isn't flying exactly; it's a sort of swimming in the air. Like being underwater, except that the medium around you isn't so dense and viscous, and you can breathe. Remember? Well, that's the feeling you get on a low-gee planetoid.

Your arms don't tend to hang at your sides, as they do on Earth or Luna, because the muscular tension tends to hold them out, just as it does in zero-gee, but there is still a definite sensation of up-and-down. If you push yourself off the floor, you tend to float in a long, slow, graceful arc, provided you don't push too hard. Magnetic soles are practically a must.

I followed the blond Mr. Feller down a series of long corridors which had been painted a pale green, which gave me the feeling that I was underwater. There were doors spaced at intervals along the corridor walls. Occasionally one of them would open and a busy looking man would cross the corridor, open another door, and disappear. From behind the doors, I could hear the drum of distant sounds.

We finally ended up in front of what looked like the only wooden door in the place. When you're çarving an office and residence out of a nickel-iron planetoid, importing wood from Earth is a purely luxury matter.

There was no name plate on that mahogany-red door; there didn't need to be.

Feller touched a thin-lined circle in the door jamb.

"You don't knock?" I asked with mock seriousness.

"No," said Feller, with a straight face. "I have to signal. Knocking wouldn't do any good. That's just wood veneer over a three-inch-thick steel slab."

The door opened and I stepped inside.

I have never seen a room quite like it. The furniture was all that same mahogany—a huge desk, nineteenth century baroque, with carved and curlicued legs; two chairs carved the same, with padded seats of maroon leather; and a chair behind the desk that might have doubled as a bishop's throne, with even fancier carving. Off to one side was a long couch upholstered in a lighter maroon. The wall-to-wall carpeting was a rich Burgundy, with a pile deep enough to run a reaper through. The walls were paneled with mahogany and hung with a couple of huge tapestries done in maroon, purple, and red. A bookcase along one wall was filled with books, every one of which had been rebound in maroon leather.

It was like walking into a cask of old claret. Or old blood.

The man sitting behind the desk looked as though he'd been built to be the lightest spot in an analogous color scheme. His suit was mauve with purple piping, and his wide, square, saggy face was florid. On his nose and cheeks, tiny lines of purple tracing made darker areas in his skin. His hair was a medium brown, but it was clipped so short that the scalp showed faintly through, and amid all that overwhelming background, even the hair looked vaguely violet.

"Come in, Mr. Oak," said Shalimar Ravenhurst.

I walked toward him across the Burgundy carpet while the blond young man discreetly closed the door behind me, leaving us alone. I didn't blame him. I was wearing a yellow union suit, and I hate to think what I must have looked like in that room.

I sat down in one of the chairs facing the desk after giving a brief shake to a thick-fingered, well-manicured, slightly oily hand.

He opened a crystal decanter that stood on one end of the desk. "Have some Madeira, Mr. Oak? Or would you like something else? I never drink spirits at this time of night."

I fought down an impulse to ask for a shot of redeye. "The Madeira will be fine, Mr. Ravenhurst."

He poured and handed me a stemmed glass nearly brimming with the wine. I joined him in an appreciative sip, then waited while he made up his mind to talk.

He leaned across the desk, looking at me with his small, dark eyes. He had an expression on his face that looked as if it were trying to sneer and leer at the same time but couldn't get much beyond the smirk stage.

"Mr. Oak, I have investigated you thoroughly—as thoroughly as it can be done, at least. My attorneys say that your reputation is A-one; that you get things done and rarely disappoint a client."

He paused as if waiting for a comment. I gave him nothing.

After a moment, he went on. "I hope that's true, Mr. Oak, because I'm going to have to trust you." He leaned back in his chair again, his eyes still on me. "Men very rarely like me, Mr. Oak. I am not a likable man. I do not pretend to be.

That's not my function." He said it as if he had said it many times before, believed it, and wished it wasn't so.

"I do not ask that you like me," he continued. "I only ask that you be loyal to my interests for the duration of this assignment." Another pause. "I have been assured by others that this will be so. I would like your assurance."

"If I take the assignment, Mr. Ravenhurst," I told him, "I'll be working for *you*. I can be bought, but once I'm bought I stay bought.

"Now, what seems to be your trouble?"

He frowned. "Well, now, let's get one thing settled: Are you working for me, or not?"

"I won't know that until I find out what the job is."

His frown deepened. "Now, see here; this is very confidential work. What happens if I tell you and you decide not to work for me?"

I sighed. "Ravenhurst, right now, you're paying me to listen to you. Even if I don't take your job, I'm going to bill you for expenses and time to come all the way out here. So, as far as listening is concerned, I'm working for you now. If I don't like the job, I'll still forget everything I'm told. All right?"

He didn't like it, but he had no choice. "All right," he said. He polished off his glass of Madeira and refilled it. My own glass was still nearly full.

"Mr. Oak," he began, "I have two problems. One is minor, the other major. But I have attempted to blow the minor problem up out of proportion, so that all the people here at Raven's Rest think that it is the only problem. They think that I brought you out here for that reason alone.

"But all that is merely cover-up for the real problem."

"Which is?" I prompted.

He leaned forward again. Apparently, it was the only exercise he ever got. "You're aware that Viking Spacecraft is one of the corporations under the management of Ravenhurst Holdings?"

I nodded. Viking Spacecraft built some of the biggest and best spacecraft in the System. It held most of Ceres—all of it, in fact, except the Government Reservation. It had moved out to the asteroids a long time back, after the big mining concerns began cutting up the smaller asteroids for metal. The raw materials are easier to come by out here than they are on Earth, and it's a devil of a lot easier to build spacecraft under low-gee conditions than it is under the pull of Earth or Luna or Mars.

"Do you know anything about the experimental robotic ships being built on Eros?" Ravenhurst asked.

"Not much," I admitted. "I've heard about them, but I don't know any of the details." That wasn't quite true, but I've found it doesn't pay to tell everybody everything you know.

"The engineering details aren't necessary," Ravenhurst said. "Besides, I don't know them, myself. The point is that Viking is trying to build a ship that will be as easy to operate as a flitterboat—a one-man cargo vessel. Perhaps even a completely automatic job for cargo, and just use a one-man crew for the passenger vessels. Imagine how that would cut the cost of transportation in the Solar System! Imagine how it would open up high-speed cargo transfer if an automatic vessel could accelerate at twenty or twenty-five gees to turnover!"

I'll give Ravenhurst this: He had a light in his eyes that showed a real excitement about the prospect he was discussing, and it wasn't due entirely to the money he might make.

"Sounds fine," I said. "What seems to be the trouble?"

His face darkened half a shade. "The company police suspect sabotage, Mr. Oak."

"How? What kind?"

"They don't know. Viking has built six ships of that type—the McGuire class, the engineers call it. Each one has been slightly different than the one before, of course, as they ironed out the bugs in their operation. But each one has been a failure. Not one of them would pass the test for space-worthiness."

"Not a failure of the drive or the ordinary mechanisms of the ship, I take it?"

Ravenhurst sniffed. "Of course not. The brain. The ships became, as you might say, *non compos mentis*. As a matter of fact, when the last one simply tried to burrow into the surface of Eros by reversing its drive, one of the roboticists said that a coroner's jury would have returned a verdict of 'suicide while of unsound mind' if there were inquests held for spaceships."

"That doesn't make much sense," I said.

"No. It doesn't. It isn't sensible. Those ships' brains shouldn't have behaved that way. Robot brains don't go mad unless they're given instructions to do so—conflicting orders, erroneous information, that sort of thing. Or, unless they have actual physical defects in the brains themselves."

"The brains can handle the job of flying a ship all right, though?" I asked. "I mean, they have the capacity for it?"

"Certainly. They're the same type that's used to control the automobile traffic on the Eastern Seaboard Highway Network of North America. If they can control the movement of millions of cars, there's no reason why they can't control a spaceship."

"No," I said, "I suppose not." I thought it over for a second, then asked, "But what do your robotics men say is causing the malfunctions?"

"That's where the problem comes in, Mr. Oak." He pursed his pudgy lips, and his eyes narrowed. "The opinions are divided. Some of the men say it's simply a case of engineering failure—that the bugs haven't been worked out of this new combination, but that as soon as they are, everything will work as smoothly as butter. Others say that only deliberate tampering could cause those failures. And still others say that there's not enough evidence to prove either of those theories is correct."

"But your opinion is that it's sabotage?"

"Exactly," said Ravenhurst, "and I know who is doing it and why."

I didn't try to conceal the little bit of surprise that gave me. "You know the man who's responsible?"

He shook his head rapidly, making his jowls wobble. "I didn't mean that. It's not a single man; it's a group."

"Maybe you'd better go into a little more detail on that, Mr. Ravenhurst."

He nodded, and this time his jowls bobbled instead of wobbled. "Some group at Viking is trying to run me out of the managerial business. They want Viking to be managed by Thurston Enterprises; they evidently think they can get a better deal from him than they can from me. If the McGuire project fails, they'll have a good chance of convincing the stock-holders that the fault lies with Ravenhurst. You follow?"

"So far," I said. "Do you think Thurston's behind this, then?"

"I don't know," he said slowly. "He might be, or he might not. If he is, that's perfectly legitimate business tactics. He's got a perfect right to try to get more business for himself if he wants to. I've undercut him a couple of times.

"But I don't think he's too deeply involved, if he's involved at all. This smacks of a personal attack against me, and I don't think that's Thurston's type of play.

"You see, things are a little touchy right now. I won't go into details, but you know what the political situation is at the moment.

"It works this way, as far as Viking is concerned: If I lose the managerial contract at Viking, a couple of my other contracts will go by the board, too—especially if it's proved that I've been lax in management or have been expending credit needlessly.

"These other two companies are actually a little shaky at the moment; I've only been managing them for a little over a year in one case and two years in the other. Their assets have come up since I took over, but they'd still dump me if they thought I was reckless."

"How can they do that?" I asked. "You have a contract, don't you?"

"Certainly. They wouldn't break it. But they'd likely ask the Government Inspectors to step in and check every step of the managerial work. Now, you and I and everybody else knows that you have to cut corners to make a business successful. If the GI's step in, that will have to stop—which means we'll show a loss heavy enough to put us out. We'll be forced to sell the contract for a pittance.

"Well, then. If Viking goes, and these other two corporations go, it'll begin to look as if Ravenhurst can't take care of himself and his companies anymore. Others will climb on the bandwagon. Contracts that are coming up for renewal will be reconsidered instead of continuing automatically. I think you can see where that would lead eventually."

I did. You don't go into the managing business these days unless you have plenty on the ball. You've got to know all the principles and all the tricks of organization and communication, and you've got to be able to waltz your way around all the roadblocks that are caused by Government laws—some of which have been floating around on the books of one nation or another for two or three centuries.

Did you know that there's a law on the American statute books that forbids the landing of a spaceship within one hundred miles of a city? That was passed back when they were using rockets, but it's never been repealed. Technically, then, it's almost impossible to land a ship anywhere on the North American continent. Long Island Spaceport is openly flouting the law, if you want to look at it that way.

A managerial combine has to know all those little things and know how to get around them. It has to be able to have the confidence of the stock-holders of a corporation—if it's run on the Western Plan—or the confidence of communal owners if it's run on the Eastern Plan.

Something like this could snowball on Ravenhurst. It isn't only the rats that desert a sinking ship; so does anyone else who has any sense.

"What I want to know, Mr. Oak," Ravenhurst continued, "is who is behind this plot, whether an individual or a group. I want to know identity and motivation."

"Is that all?" I eyed him skeptically.

"No. Of course not. I want you to make sure that the MG-YR-7 isn't sabotaged. I want you to make sure it's protected from whatever kind of monkey wrenches are being thrown into its works."

"It's nearly ready for testing now, isn't it?" I asked.

"It is ready. It seems to be in perfect condition so far. Viking is already looking for a test pilot. It's still in working order now, and I want to be certain that it will remain so."

I cocked my head to one side and gave him my Interrogative And Suspicious Glance—Number 9 in the manual. "You didn't do any checking on the first six McGuire ships. You wait until this one is done before calling me. Why the delay, Ravenhurst?"

It didn't faze him. "I became suspicious after McGuire 6 failed. I put Colonel Brock on it."

I nodded. I'd had dealings with Brock. He was head of Ravenhurst's Security Guard. "Brock didn't get anywhere," I said.

"He did not. His own face is too well known for him to have investigated personally, and he's not enough of an actor to get away with using a plexiskin mask. He had to use underlings. And I'm afraid some of them might be in the pay of the ... ah ... opposition. They got nowhere."

"In other words, you may have spies in your own organization who are working with the Viking group. Very interesting. That means they know I'm working for you, which will effectively seal me up, too. You might as well have kept Brock on the job."

He smiled in a smug, superior sort of way that some men might have resented. I did. Even though I'd fed him the line so that he could feel superior, knowing that a smart operator like Ravenhurst would already have covered his tracks. I couldn't help wishing I'd told him simply to trot out his cover story instead of letting him think I believed it had never occurred to either of us before.

"As far as my staff knows, Mr. Oak, you are here to escort my daughter, Jaqueline, to Braunsville, Luna. You will, naturally, have to take her to Ceres in your flitterboat, where you will wait for a specially chartered ship to take you both to Luna. That will be a week after you arrive. Since the McGuire 7 is to be tested within three days, that should give you ample time."

"If it doesn't?"

"We will consider that possibility if and when it becomes probable. I have a great deal of faith in you."

"Thanks. One more thing: why do you think anybody will swallow the idea that your daughter needs a private bodyguard to escort her to Braunsville?"

His smile broadened a little. "You have not met my daughter, Mr. Oak. Jaqueline takes after me in a great many respects, not the least of which is her desire to have things her own way and submit to no man's yoke, as the saying goes. I have had a difficult time with her, sir; a difficult time. It is and has been a matter of steering a narrow course between the Scylla of breaking her spirit with too much discipline and the Charybdis of allowing her to ruin her life by letting her go hog wild. She is seventeen now, and the time has come to send her to a school where she will receive an education suitable to her potentialities and abilities, and discipline which will be suitable to her spirit.

"Your job, Mr. Oak, will be to make sure she gets there. You are not a bodyguard in the sense that you must protect her from the people around her. Quite the contrary, *they* may need protection from *her*. You are to make sure she arrives in Braunsville on schedule. She is perfectly capable of taking it in her head to go scooting off to Earth if you turn your back on her."

Still smiling, he refilled his glass. "Do have some more Madeira, Mr. Oak. It's really an excellent year."

I let him refill my glass.

"That, I think, will cover your real activities well enough. My daughter will, of course, take a tour of the plant on Ceres, which will allow you to do whatever work is necessary."

He smiled at me.

I didn't smile back.

"Up till now, this sounded like a pretty nice assignment," I said. "But I don't want it now. I can't take care of a teenage girl with a desire for the bright lights of Earth while I investigate a sabotage case."

I knew he had an out; I was just prodding him into springing it.

He did. "Of course not. My daughter is not as scatterbrained as I have painted her. She is going to help you."

"*Help* me?"

"Exactly. You are ostensibly her bodyguard. If she turns up missing, you will, of course, leave no stone unturned to find her." He chuckled. "And Ceres is a fairly large stone."

I thought it over. I still didn't like it too well, but if Jaqueline wasn't going to be too much trouble to take care of, it might work out. And if she did get to be too much trouble, I could see to it that she was unofficially detained for a while.

"All right, Mr. Ravenhurst," I said, "you've got yourself a man for both jobs."

"Both?"

"I find out who is trying to sabotage the McGuire ship, and I baby-sit for you. That's two jobs. And you're going to pay for both of them."

"I expected to," said Shalimar Ravenhurst.

Fifteen minutes later, I was walking into the room where I'd left my vac suit. There was a girl waiting for me.

She was already dressed in her vac suit, so there was no way to be sure, but she looked as if she had a nice figure underneath the suit. Her face was rather unexceptionally pretty, a sort of nice-girl-next-door face. Her hair was a reddish brown and was cut fairly close to the skull; only a woman who never intends to be in a vac suit in free fall can afford to let her hair grow.

"Miss Ravenhurst?" I asked.

She grinned and stuck out a hand. "Just call me Jack. And I'll call you Dan. O.K.?"

I grinned and shook her hand because there wasn't much else I could do. Now I'd met the Ravenhursts: A father called Shalimar and a daughter called Jack.

And a spaceship named McGuire.

I gave the flitterboat all the push it would take to get us to Ceres as fast as possible. I don't like riding in the things. You sit there inside a transite hull, which has two bucket seats inside it, fore and aft, astraddle the drive tube, and you guide from one beacon to the next while you keep tabs on orbital positions by radio. It's a long jump from one rock to the next, even in the asteroid belt, and you have to live inside your vac suit until you come to a stopping place where you can spend an hour or so resting before you go on. It's like driving cross-continent in an automobile, except that the signposts and landmarks are constantly shifting position. An inexperienced man can get lost easily in the Belt.

I was happy to find that Jack Ravenhurst knew how to handle a flitterboat and could sight navigate by the stars. That meant that I could sleep while she piloted and vice-versa. The trip back was a lot easier and faster than the trip out had been.

I was glad, in a way, that Ceres was within flitterboat range of Raven's Rest. I don't like the time wasted in waiting for a regular spaceship, which you have to do when your target is a quarter of the way around the Belt from you. The cross-system jumps don't take long, but getting to a ship takes time.

The Ravenhurst girl wasn't much of a talker while we were en route. A little general chitchat once in a while, then she'd clam up to do a little mental orbit figuring. I didn't mind. I was in no mood to pump her just yet, and I was usually figuring orbits myself. You get in the habit after a while.

When the Ceres beacon came into view, I was snoozing. Jack reached forward and shook my shoulder. "Decelerating toward Ceres," she said. "Want to take over from here on?" Her voice sounded tinny and tired in the earphones of my fishbowl.

"O.K.; I'll take her in. Have you called Ceres Field yet?"

"Not yet. I figured that you'd better do that, since it's your flitterboat."

I said O.K. and called Ceres. They gave me a traffic orbit, and I followed it in to Ceres Field.

It was a lot bigger than the postage-stamp field on Raven's Rest, and more brightly lit, and a lot busier, but it was basically the same idea—a broad, wide, smooth area that had been carved out of the surface of the nickel-iron with a focused sun beam. One end of it was reserved for flitterboats; three big spaceships sat on the other end, looking very *noblesse oblige* at the little flitterboats.

I clamped down, gave the key to one of the men behind the desk after we had gone below, and turned to Jack. "I suggest we go to the hotel first and get a shower and a little rest. We can go out to Viking tomorrow."

She glanced at her watch. Like every other watch and clock in the Belt, it was set for Greenwich Standard Time. What's the point in having time zones in space?

"I'm not tired," she said brightly. "I got plenty of sleep while we were on the way. Why don't we go out tonight? They've got a bounce-dance place called *Bali*'s that—"

I held up a hand. "No. You may not be tired, but I am. Remember, I went all the way out there by myself, and then came right back.

"I need at least six hours sleep in a nice, comfortable bed before I'll be able to move again."

The look she gave me made me feel every one of my thirty-five years, but I didn't intend to let her go roaming around at this stage of the game.

Instead, I put her aboard one of the little rail cars, and we headed for the Viking Arms, generally considered the best hotel on Ceres.

Ceres has a pretty respectable gee pull for a planetoid: Three per cent of Standard. I weigh a good, hefty five pounds on the surface. That makes it a lot easier to walk around on Ceres than on, say, Raven's Rest. Even so, you always get the impression that one of the little rail cars that scoots along the corridors is climbing uphill all the way, because the acceleration is greater than any measly thirty centimeters per second squared.

Jack didn't say another word until we reached the Viking, where Ravenhurst had thoughtfully made reservations for adjoining rooms. Then, after we'd registered, she said: "We could at least get something to eat."

"That's not a bad idea. We can get something to line our stomachs, anyway. Steak?"

She beamed up at me. "Steak. Sounds wonderful after all those mushy concentrates. Let's go."

The restaurant off the lobby was just like the lobby and the corridors outside—a big room hollowed out of the metal of the asteroid. The walls had been painted to prevent rusting, but they still bore the roughness left by the sun beam that had burnt them out.

We sat down at a table, and a waiter brought over a menu. The place wouldn't be classed higher than a third-rate cafe on Earth, but on Ceres it's considered one of the better places. The prices certainly compare well with those of the best New York or Moscow restaurants, and the price of meat, which has to be shipped from Earth, is—you should pardon the gag—astronomical.

That didn't bother me. Steaks for two would go right on the expense account. I mentally thanked Mr. Ravenhurst for the fine slab of beef when the waiter finally brought it.

While we were waiting, though, I lit a cigarette and said: "You're awfully quiet, Jack."

"Am I? Men are funny."

"Is that meant as a conversational gambit, or an honest observation?"

"Observation. I mean, men are always complaining that girls talk too much, but if a girl keeps her mouth shut, they think there's something wrong with her."

"Uh-huh. And you think that's a paradox or something?"

She looked puzzled. "Isn't it?"

"Not at all. The noise a jackhammer makes isn't pleasant at all, but if it doesn't make that noise, you figure it isn't functioning properly. So you wonder why."

Out of the corner of my eye, I had noticed a man wearing the black-and-gold union suit of Ravenhurst's Security Guard coming toward us from the door, using the gliding shuffle that works best under low gee. I ignored him to listen to Jack Ravenhurst.

"That has all the earmarks of a dirty crack," she said. The tone of her voice indicated that she wasn't sure whether to be angry or to laugh.

"Hello, Miss Ravenhurst; Hi, Oak." Colonel Brock had reached the table. He stood there, smiling his rather flat smile, while his eyes looked us both over carefully.

He was five feet ten, an inch shorter than I am, and lean almost to the point of emaciation. His scarred, hard-bitten face looked as though it had gotten that way when he tried to kiss a crocodile.

"Hello, Brock," I said. "What's new?"

Jack gave him a meaningless smile and said: "Hello, colonel." She was obviously not very impressed with either of us.

"Mind if I sit?" Brock asked.

We didn't, so he sat.

"I'm sorry I missed you at the spaceport," Brock said seriously, "but I had several of my boys there with their eyes open." He was quite obviously addressing Jack, not me.

"It's all right," Jack said. "I'm not going anywhere this time." She looked at me and gave me an odd grin. "I'm going to stay home and be a good girl this time around."

Colonel Brock's good-natured chuckle sounded about as genuine as the ring of a lead nickel. "Oh, you're no trouble, Miss Ravenhurst."

"Thank you, kind sir; you're a poor liar." She stood up and smiled sweetly. "Will you gentlemen excuse me a moment?"

We would and did. Colonel Brock and I watched her cross the room and disappear through a door. Then he turned to look at me, giving me a wry grin and shaking his head a little sadly. "So you got saddled with Jack the Ripper, eh, Oak?"

"Is she that bad?"

His chuckle was harsher this time, and had the ring of truth. "You'll find out. Oh, I don't mean she's got the morals of a cat or anything like that. So far as I know, she's still waiting for Mister Right to come along."

"Drugs?" I asked. "Liquor?"

"A few drinks now and then—nothing else," Brock said. "No, it's none of the usual things. It isn't what *she* does that counts; it's what she talks other people into doing. She's a convincer."

"That sounds impressive," I said. "What does it mean?"

His hard face looked wolfish, "I ought to let you find out for yourself. But, no; that wouldn't be professional courtesy, and it wouldn't be ethical."

"Brock," I said tiredly, "I have been given more runarounds in the past week than Mercury has had in the past millennium. I expect clients to be cagey, to hold back information, and to lie. But I didn't expect it of you. Give."

He nodded brusquely. "As I said, she's a convincer. A talker. She can talk people into doing almost anything she wants them to."

"For instance?"

"Like, for instance, getting all the patrons at the *Bali* to do a snake dance around the corridors in the altogether. The Ceres police broke it up, but she was nowhere to be found."

He said it so innocently that I knew he'd been the one to get her out of the mess.

"And the time," he continued, "that she almost succeeded in getting a welder named Plotkin elected Hereditary Czar of Ceres. She'd have succeeded, too, if she hadn't made the mistake of getting Plotkin himself up to speak in front of his loyal supporters. After that, everybody felt so silly that the movement fell apart."

He went on, reciting half a dozen more instances of the girl's ability to influence people without winning friends. None of them were new to me; they were all on file in the Political Survey Division of the United Nations Government on Earth, plus several more which Colonel Brock either neglected to tell me or wasn't aware of himself.

But I listened with interest; after all, I wasn't supposed to know any of these things. I am just a plain, ordinary, "confidential expediter". That's what it says on the door of my office in New York, and that's what it says on my license. All very legal and very dishonest.

The Political Survey Division is very legal and very dishonest, too. Theoretically, it is supposed to be nothing but a branch of the System Census Bureau; it is supposed to do nothing but observe and tabulate political trends. The actual fact that it is the Secret Service branch of the United Nations Government is known only to relatively few people.

I know it because I work for the Political Survey Division.

The PSD already had men investigating both Ravenhurst and Thurston, but when they found out that Ravenhurst was looking for a confidential expediter, for a special job, they'd shoved me in fast.

It isn't easy to fool sharp operators like Colonel Brock, but, so far, I'd been lucky enough to get away with it by playing ignorant-but-not-stupid.

The steaks were brought, and I mentally saluted Ravenhurst, as I had promised myself I would. Then I rather belatedly asked the colonel if he'd eat with us.

"No," he said, with a shake of his head. "No, thanks. I've got to get things ready for her visit to the Viking plant tomorrow."

"Oh? Hiding something?" I asked blandly.

He didn't even bother to look insulted. "No. Just have to make sure she doesn't get hurt by any of the machinery, that's all. Most of the stuff is automatic, and she has a habit of getting too close. I guess she thinks she can talk a machine out of hurting her as easily as she can talk a man into standing on his head."

Jack Ravenhurst was coming back to the table. I noticed that she'd fixed her hair nicely and put on make-up. It made her look a lot more feminine than she had while she was on the flitterboat.

"Well," she said as she sat down, "have you two decided what to do with me?"

Colonel Brock just smiled and said: "I guess we'll have to leave that up to you, Miss Ravenhurst." Then he stood up. "Now, if you'll excuse me, I'll be about my business."

Jack nodded, gave him a quick smile, and fell to on her steak with the voraciousness of an unfed chicken in a wheat bin.

Miss Jaqueline Ravenhurst evidently had no desire to talk to me at the moment.

On Ceres, as on most of the major planetoids, a man's home is his castle, even if it's only a hotel room. Raw nickel-iron, the basic building material, is so cheap that walls and doors are seldom made of anything else, so a hotel room is more like a vault than anything else on Earth. Every time I go into one of the hotels on Ceres or Eros, I get the feeling that I'm either a bundle of gold certificates or a particularly obstreperous prisoner being led to a medieval solitary confinement cell. They're not pretty, but they're *solid.*

Jack Ravenhurst went into her own room after flashing me a rather hurt smile that was supposed to indicate her disappointment in not being allowed to go nightclubbing. I gave her a big-brotherly pat on the shoulder and told her to get plenty of sleep, since we had to be up bright and early in the morning.

Once inside my own room, I checked over my luggage carefully. It had been brought there from the spaceport, where I'd checked it before going to Ravenhurst's Raven's Rest, on orders from Ravenhurst himself. This was one of several rooms that Ravenhurst kept permanently rented for his own uses, and I knew that Jack kept a complete wardrobe in her own rooms.

There were no bugs in my luggage—neither sound nor sight spying devices of any kind. Not that I would have worried if there had been; I just wanted to see if anyone was crude enough to try that method of smuggling a bug into the apartment.

The door chime pinged solemnly.

I took a peek through the door camera and saw a man in a bellboy's uniform, holding a large traveling case. I recognized the face, so I let him in.

"The rest of your luggage, sir," he said with a straight face.

"Thank you very much," I told him. I handed him a tip, and he popped off.

This stuff was special equipment that I hadn't wanted Ravenhurst or anybody else to get his paws into.

I opened it carefully with the special key, slid a hand under the clothing that lay on top for camouflage, and palmed the little detector I needed. Then I went around the room, whistling gently to myself.

The nice thing about an all-metal room is that it's impossible to hide a self-contained bug in it that will be of any use. A small, concealed broadcaster can't broadcast any farther than the walls, so any bug has to have wires leading out of the room.

I didn't find a thing. Either Ravenhurst kept the room clean or somebody was using more sophisticated bugs than any I knew about. I opened the traveling case again and took out one of my favorite gadgets. It's a simple thing, really: a noise generator. But the noise it generates is non-random noise. Against a background of "white," purely random noise, it is possible to pick out a conversation, even if the conversation is below the noise level, simply because conversation is patterned. But this little generator of mine was non-random. It was the multiple recording of ten thousand different conversations, all meaningless, against a background of "white" noise. Try that one on your differential analyzers.

By the time I got through, nobody could tap a dialogue in that room, barring, as I said, bugs more sophisticated than any the United Nations knew about.

Then I went over and tapped on the communicating door between my room and Jack Ravenhurst's. There was no answer.

I said, "Jack, I'm coming in. I have a key."

She said, "Go away. I'm not dressed. I'm going to bed."

"Grab something quick," I told her. "I'm coming in."

I keyed open the door.

She was no more dressed for bed than I was, unless she made a habit of sleeping in her best evening togs. Anger blazed in her eyes for a second, then that faded, and she tried to look all sweetness and light.

"I was trying on some new clothes," she said innocently.

A lot of people might have believed her. The emotional field she threw out, encouraging utter belief in her every word, was as powerful as any I'd ever felt. I just let it wash past me and said: "Come into my room for a few minutes, Jack; I want to talk to you."

I didn't put any particular emphasis into it. I don't have to. She came.

Once we were both inside my shielded room with the walls vibrating with ten thousand voices and a hush area in the center, I said patiently, "Jack, I personally don't care where you go or what you do. Tomorrow, you can do your vanishing act and have yourself a ball, for all I care. But there are certain things that have to be done first. Now, sit down and listen."

She sat down, her eyes wide. Evidently, nobody had ever beaten her at her own game before.

"Tonight, you'll stay here and get some sleep. Tomorrow, we go for a tour of Viking, first thing in the morning. Tomorrow afternoon, as soon as I think the time is ripe, you can sneak off. I'll show you how to change your appearance so you won't be recognized. You can have all the fun you want for twenty-four hours. I, of course, will be hunting high and low for you, but I won't find you until I have finished my investigation.

"On the other hand, I want to know where you are at all times, so that I can get in touch with you if I need you. So, no matter where you are, you'll keep in touch by phoning BANning 6226 every time you change location. Got that number?"

She nodded. "BANning 6226," she repeated.

"Fine. Now, Brock's agents will be watching you, so I'll have to figure out a way to get you away from them, but that won't be too hard. I'll let you know at the proper time. Meanwhile, get back in there, get ready for bed, and get some sleep. You'll need it. Move."

She nodded rather dazedly, got up, and went to the door. She turned, said goodnight in a low, puzzled voice, and closed the door.

Half an hour later, I quietly sneaked into her room just to check. She was sound asleep in bed. I went back to my own room, and got some sack time myself.

"It's a pleasure to have you here again, Miss Ravenhurst," said Chief Engineer Midguard. "Anything in particular you want to see this time?" He said it as though he actually enjoyed taking the boss' teenage daughter through a spacecraft plant.

Maybe he did, at that. He was a paunchy, graying man in his sixties, who had probably been a rather handsome lady-killer for the first half-century of his life, but he was approaching middle age now, which has a predictable effect on the telly-idol type.

Jack Ravenhurst was at her regal best, with the kind of *noblesse oblige* that would bring worshipful gratitude to the heart of any underling. "Oh, just a quick run-through on whatever you think would be interesting, Mr. Midguard; I don't want to take up too much of your time."

Midguard allowed as how he had a few interesting things to show her, and the party, which also included the watchful and taciturn Colonel Brock, began to make the rounds of the Viking plant.

There were three ships under construction at the time: two cargo vessels and a good-sized passenger job. Midguard seemed to think that every step of spacecraft construction was utterly fascinating—for which, bully for him—but it was pretty much of a drag as far as I was concerned. It took three hours.

Finally, he said, "Would you like to see the McGuire-7?"

Why, yes, of course she would. So we toddled off to the new ship while Midguard kept up a steady line of patter.

"We think we have all the computer errors out of this one, Miss Ravenhurst. A matter of new controls and safety devices. We feel that the trouble with the first six machines was that they were designed to be operated by voice orders by any qualified human operator. The trouble is that they had no way of telling just who was qualified. The brains are perfectly capable of distinguishing one individual from another, but they can't tell whether a given individual is a space pilot or a janitor. In fact—"

I marked the salient points in his speech. The MG-YR-7 would be strictly a one-man ship. It had a built-in dog attitude—friendly toward all humans, but loyal only to its master. Of course, it was likely that the ship would outlast its master, so its loyalties could be changed, but only by the use of special switching keys.

The robotics boys still weren't sure why the first six had gone insane, but they were fairly certain that the primary cause was the matter of too many masters. The brilliant biophysicist, Asenion, who promulgated the Three Laws of Robotics in the last century, had shown in his writings that they were unattainable ideals—that they only told what a perfect robot *should* be, not what a robot actually was.

The First Law, for instance, would forbid a robot to harm a human being, either by action or inaction. But, as Asenion showed, a robot could be faced with a situation which allowed for only two possible decisions, both of which required that a human being be harmed. In such a case, the robot goes insane.

I found myself speculating what sort of situation, what sort of Asenion paradox, had confronted those first six ships. And whether it had been by accident or design. Not that the McGuire robots had been built in strict accord with the Laws of Robotics; that was impossible on the face of it. But no matter how a perfectly logical machine is built, the human mind can figure out a way to goof it up because the human mind is capable of transcending logic.

The McGuire ship was a little beauty. A nice, sleek, needle, capable of atmospheric as well as spatial navigation, with a mirror-polished, beryl-blue surface all over the sixty-five feet of her—or his?—length.

It was standing upright on the surface of the planetoid, a shining needle in the shifting sunlight, limned against the star-filled darkness of space. We looked at it through the transparent viewport, and then took the flexible tube that led to the air lock of the ship.

The ship was just as beautiful inside as it was outside. Neat, compact, and efficient. The control room—if such it could be called—was like no control room I'd ever seen before. Just an acceleration couch and observation instruments. Midguard explained that it wasn't necessary to be a pilot to run the ship; any person who knew a smattering of astronavigation could get to his destination by simply telling the ship what he wanted to do.

Jack Ravenhurst took in the whole thing with wide-eyed interest.

"Is the brain activated, Mr. Midguard?" she asked.

"Oh, yes. We've been educating him for the past month, pumping information in as rapidly as he could record it and index it. He's finished with that stage now; we're just waiting for the selection of a test pilot for the final shakedown cruise." He was looking warily at Jack as he spoke, as if he were waiting for something.

Evidently, he knew what was coming. "I'd like to talk to him," Jack said. "It's so interesting to carry on an intelligent conversation with a machine."

"I'm afraid that's impossible, Miss Ravenhurst," Midguard said rather worriedly. "You see, McGuire's primed so that the first man's voice he hears will be identified as his master. It's what we call the 'chick reaction'. You know: the first moving thing a newly-hatched bird sees is regarded as the mother, and, once implanted, that order can't be rescinded. We can change McGuire's orientation in that respect, but we'd rather not have to go through that. After the test pilot establishes contact, you can talk to him all you want."

"When will the test pilot be here?" Jack asked, still as sweet as sucrodyne.

"Within a few days. It looks as though a man named Nels Bjornsen will be our choice. You may have heard of him."

"No," she said, "but I'm sure your choice will be correct."

Midguard still felt apologetic. "Well, you know how it is, Miss Ravenhurst; we can't turn a delicate machine like this over to just anyone for the first trial. He has to be a man of good judgment and fast reflexes. He has to know exactly what to say and when to say it, if you follow me."

"Oh, certainly; certainly." She paused and looked thoughtful. "I presume you've taken precautions against anyone stealing in here and taking control of the ship."

Midguard smiled and nodded wisely. "Certainly. Communication with McGuire can't be established unless and until two keys are used in the activating panel. I carry one; Colonel Brock has the other. Neither of us will give his key up to anyone but the accredited test pilot. And McGuire himself will scream out an alarm if anyone tries to jimmy the locks. He's his own burglar alarm."

She nodded. "I see." A pause. "Well, Mr. Midguard, I think you've done a very commendable job. Thank you so much. Is there anything else you feel I should see?"

"Well—" He was smilingly hesitant. "If there's anything else you want to see, I'll be glad to show it to you. But you've already seen our ... ah ... *piece de resistance*, so to speak."

She glanced at her wrist. It had been over four hours since we'd started. "I am rather tired," Jack said. "And hungry, too. Let's call it a day and go get something to eat."

"Fine! Fine!" Midguard said. "I'll be honored to be your host, if I may. We could have a little something at my apartment."

I knew perfectly well that he'd had a full lunch prepared and waiting.

The girl acknowledged his invitation and accepted it. Brock and I trailed along like the bodyguards we were supposed to be. I wondered whether or not Brock suspected me of being more than I appeared to be. If he didn't, he was stupider than I thought; on the other hand, he could never be sure. I wasn't worried about his finding out that I was a United Nations agent; that was a pretty remote chance. Brock didn't even know the United Nations Government *had* a Secret Service; it was unlikely that he would suspect me of being an agent of a presumably nonexistent body.

But he could very easily suspect that I had been sent to check on him and the Thurston menace, and, if he had any sense, he actually did. I wasn't going to give him any verification of that suspicion if I could help it.

Midguard had an apartment in the executive territory of the Viking reservation, a fairly large place with plastic-lined walls instead of the usual painted nickel-iron. Very luxurious for Ceres.

The meal was served with an air of subdued pretension that made everybody a little stiff and uncomfortable, with the possible exception of Jack Ravenhurst, and the definite exception of myself. I just listened politely to the strained courtesy that passed for small talk and waited for the chance I knew would come at this meal.

After the eating was all over, and we were all sitting around with cigarettes going and wine in our glasses, I gave the girl the signal we had agreed upon. She excused herself very prettily and left the room.

After fifteen minutes, I began to look a little worried. The bathroom was only a room away—we were in a dining area, and the bathroom was just off the main bedroom—and it shouldn't have taken her that long to brush her hair and powder her face.

I casually mentioned it to Colonel Brock, and he smiled a little.

"Don't worry, Oak; even if she does walk out of this apartment, my men will be following her wherever she goes. I'd have a report within one minute after she left."

I nodded, apparently satisfied. "I've been relying on that," I said. "Otherwise, I'd have followed her to the door."

He chuckled and looked pleased.

Ten minutes after that, even he was beginning to look a little worried. "Maybe we'd better go check," he said. "She might have hurt herself or ... or become ill."

Midguard looked flustered. "Now, just a minute, colonel! I can't allow you to just barge in on a young girl in the ... ah ... bathroom. Especially not Miss Ravenhurst."

Brock made his decision fast; I'll give him credit for that.

"Get Miss Pangloss on the phone!" he snapped. "She's just down the corridor. She'll come down on your orders."

At the same time, he got to his feet and made a long jump for the door. He grabbed the doorpost as he went by, swung himself in a new orbit, and launched himself toward the front door. "Knock on the bathroom door, Oak!" he bawled as he left.

I did a long, low, flat dive toward the bedroom, swung left, and brought myself up sharply next to the bathroom door. I pounded on the door. "Miss Ravenhurst! Jack! Are you all right?"

No answer.

Good. There shouldn't have been.

Colonel Brock fired himself into the room and braked himself against the wall. "Any answer?"

"No."

"My men outside say she hasn't left." He rapped sharply on the door with the butt of his stun gun. "Miss Ravenhurst! Is there anything the matter?"

Again, no answer.

I could see that Brock was debating on whether he should go ahead and charge in by himself without waiting for the female executive who lived down the way. He was still debating when the woman showed up, escorted by a couple of the colonel's uniformed guards.

Miss Pangloss was one of those brisk, efficient, middle-aged career-women who had no fuss or frills about her. She had seen us knocking on the door, so she didn't bother to do any knocking herself. She just opened the door and went in.

The bathroom was empty.

Again, as it should be.

All hell broke loose then, with me and Brock making most of the blather. It took us nearly ten minutes to find that the only person who had left the area had been an elderly, thin man who had been wearing the baggy protective clothing of a maintenance man.

By that time, Jack Ravenhurst had been gone more than forty minutes. She could be almost anywhere on Ceres.

Colonel Brock was furious and so was I. I sneered openly at his assurance that the girl couldn't leave and then got sneered back at for letting other people do what was supposed to be my job. That phase only lasted for about a minute, though.

Then Colonel Brock muttered: "She must have had a plexiskin mask and a wig and the maintenance clothing in her purse. As I recall, it was a fairly good-sized one." He didn't say a word about how careless I had been to let her put such stuff in her purse. "All right," he went on, "we'll find her."

"I'm going to look around, too," I said. "I'll keep in touch with your office." I got out of there.

I got to a public phone as fast as I could, punched BANning 6226, and said: "Marty? Any word?"

"Not yet."

"I'll call back."

I hung up and scooted out of there.

I spent the next several hours pushing my weight around all over Ceres. As the personal representative of Shalimar Ravenhurst, who was manager of Viking Spacecraft, which was, in turn, the owner of Ceres, I had a lot of weight to push around. I had every executive on the planetoid jumping before I was through.

Colonel Brock, of course, was broiling in his own juices. He managed to get hold of me by phone once, by calling a Dr. Perelson whom I was interviewing at the time.

The phone chimed, Perelson said, "Excuse me," and went to answer. I could hear his voice from the other room.

"Mr. Daniel Oak? Yes; he's here. Well, yes. Oh, all sorts of questions, colonel." Perelson's voice was both irritated and worried. "He says Miss Ravenhurst is missing; is that so? Oh? Well, does this man have any right to question me this way? Asking me? About everything!... How well I know the girl, the last time I saw her—things like that. Good heavens, we've hardly met!" He was getting exasperated now. "But does he have the authority to ask these questions? Oh. Yes. Well, of course, I'll be glad to co-operate in any manner I can ... Yes ... Yes. All right, I'll call him."

I got up from the half-reclining angle I'd been making with the wall, and shuffled across the room as Dr. Perelson stuck his head around the corner and said, "It's for you." He looked as though someone had put aluminum hydrogen sulfate in his mouthwash.

I picked up the receiver and looked at Brock's face in the screen. He didn't even give me a chance to talk. "What are you trying to do?" he shouted explosively.

"Trying to find Jaqueline Ravenhurst," I said, as calmly as I could.

"Oak, you're a maniac! Why, by this time, it's all over Ceres that the boss' daughter is missing! Shalimar Ravenhurst will have your hide for this!"

"He will?" I gave him Number 2—the wide-eyed innocent stare. "Why?"

"Why, you idiot, I thought you had sense enough to know that this should be kept quiet! She's pulled this stunt before, and we always managed to quiet things down before anything happened! We've managed to keep everything under cover and out of the public eye ever since she was fifteen, and now you blow it all up out of proportion and create a furore that won't ever be forgotten!"

He gave his speech as though it had been written for him in full caps, with three exclamation points after every sentence, and added gestures and grimaces after every word.

"Just doing what I thought was best," I said. "I want to find her as soon as possible."

"Well, stop it! Now! Let us handle it from here on in!"

Then I lowered the boom. "Now *you* listen, Brock. I am in charge of Jack Ravenhurst, not you. I've lost her, and I'll find her. I'll welcome your co-operation, and I'd hate to have to fight you, but if you don't like the way I'm handling it, you can just tell your boys to go back to their regular work and let me handle it alone, without interference. Now, which'll it be?"

He opened his mouth, closed it, and blew out his breath from between his lips. Then he said: "All right. The damage has been done, anyhow. But don't think I won't report all this to Ravenhurst as soon as I can get a beam to Raven's Rest."

"That's your job and your worry, not mine. Now, have you got any leads?"

"None," he admitted.

"Then I'll go out and dig up some. I'll let you know if I need you." And I cut off.

Dr. Perelson was sitting on his couch, with an expression that indicated that the pH of his saliva was hovering around one point five.

I said, "That will be all, Dr. Perelson. Thank you for your co-operation." And I walked out into the corridor, leaving him with a baffled look.

At the next public phone, I dialed the BANning number again.

"Any news?"

"Not from her; she hasn't reported in at all."

"I didn't figure she would. What else?"

"Just as you said," he told me. "With some cute frills around the edges. Ten minutes ago, a crowd of kids—sixteen to twenty-two age range—about forty of 'em—started a songfest and football game in the corridor outside Colonel Brock's place. The boys he had on duty there recognized the Jack Ravenhurst touch, and tried to find her in the crowd. Nothing doing. Not a sign of her."

"That girl's not only got power," I said, "but she's bright as a solar flare."

"Agreed. She's headed up toward Dr. Midguard's place now. I don't know what she has in mind, but it ought to be fun to watch."

"Where's Midguard now?" I asked.

"Hovering around Brock, as we figured. He's worried and feels responsible because she disappeared from his apartment, as predicted."

"Well, I've stirred up enough fuss in this free-falling anthill to give them all the worries they need. Tell me what's the overall effect?"

"Close to perfect. It's slightly scandalous and very mysterious, so everybody's keeping an eye peeled. If anyone sees Jaqueline Ravenhurst, they'll run to a phone, and naturally she's been spotted by a dozen different people in a dozen different places already.

"You've got both Brock's Company guards and the civil police tied up for a while."

"Fine. But be sure you keep the boys who are on her tail shifting around often enough so that she doesn't spot them."

"Don't worry your thick little head about that, Dan," he said. "They know their business. Are you afraid they'll lose her?"

"No, I'm not, and you know it. I just don't want her to know she's being followed. If she can't ditch her shadow, she's likely to try to talk to him and pull out all the stops convincing him that he should go away."

"You think she could? With *my* boys?"

"No, but if she tries it, it'll mean she knows she's being followed. That'll make it tougher to keep a man on her trail. Besides, I don't want her to try to convince him and fail."

"*Ich graben Sie.* On the off chance that she does spot one and gives him a good talking to, I'll pass along the word that the victim is to walk away meekly and get lost."

"Good," I said, "but I'd rather she didn't know."

"She won't. You're getting touchy, Dan; 'pears to me you'd rather be doing that job yourself, and think nobody can handle it but you."

I gave him my best grin. "You are closer than you know. O.K., I'll lay off. You handle your end of it and I'll handle mine."

"A fair exchange is no bargain. Go, and sin no more."

"I'll buzz you back before I go in," I said, and hung up.

Playing games inside a crowded asteroid is not the same as playing games in, say, Honolulu or Vladivostok, especially when that game is a combination of hide-and-seek and ring-around-the-Rosie. The trouble is lack of communication. Radio contact is strictly line-of-sight inside a hunk of metal. Radar beams can get a little farther, but a man has to be an expert billiards player to bank a reflecting beam around very many corners, and even that would depend upon the corridors being empty, which they never are. To change the game analogy again, it would be like trying to sink a ninety-foot putt across Times Square on New Year's Eve.

Following somebody isn't anywhere near as easy as popular fiction might lead you to believe. Putting a tail on someone whose spouse wants divorce evidence is relatively easy, but even the best detectives can lose a man by pure mischance. If the tailee, for instance, walks into a crowded elevator and the automatic computer decides that the car is filled to the limit, the man who's tailing him will be left facing a closed door. Something like that can happen by accident, without any design on the part of the tailee.

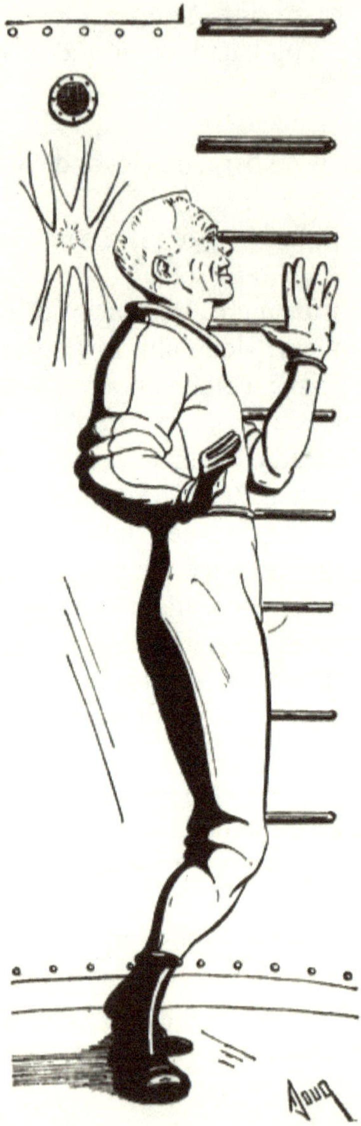

If you use a large squad of agents, all in radio contact with one another, that kind of loss can be reduced to near zero by simply having a man covering every possible escape route.

But if the tailee knows, or even suspects, that he's being followed, wants to get away from his tail, and has the ability to reason moderately well, it requires an impossibly large team to keep him in sight. And if that team has no fast medium of communication, they're licked at the onset.

In this case, we were fairly certain of Jack Ravenhurst's future actions, and so far our prophecies had been correct ... but if she decided to shake her shadows, fun would be had by all.

And as long as I had to depend on someone else to do my work for me, I was going to be just the teenchiest bit concerned about whether they were doing it properly.

I decided it was time to do my best to imitate a cosmic-ray particle, and put on a little speed through the corridors that ran through the subsurface of Ceres.

My vac suit was in my hotel room. One of the other agents had picked it up from my flitterboat and packed it carefully into a small attaché case. I'd planned my circuit so that I'd be near the hotel when things came to the proper boil, so I did a lot of diving, breaking all kinds of traffic regulations in the process.

I went to my room, grabbed the attaché case, checked it over quickly—never trust another man to check your vac suit for you—and headed for the surface.

Nobody paid any attention to me when I walked out of the air lock onto the spacefield. There were plenty of people moving in and out, going to and from their ships and boats. It wasn't until I reached the edge of the field that I realized that I had over-played my hand with Colonel Brock. It was only by the narrowest hair, but that had been enough to foul up my plans. There were guards surrounding the perimeter with radar search beams.

As I approached, one of the guards walked toward me and made a series of gestures with his left hand—two fingers up, fist, two fingers up, fist, three fingers up. I set my suit phone for 223; the guy's right hand was on the butt of his stun gun.

"Sorry, sir," came his voice. "We can't allow anyone to cross the field perimeter. Emergency."

"My name's Oak," I said tiredly. "Daniel Oak. What is going on here?"

He came closer and peered at me. Then: "Oh, yes, sir; I recognize you. We're ... uh—" He waved an arm around. "Uh ... looking for Miss Ravenhurst." His voice lowered conspiratorially. I could tell that he was used to handling the Ravenhurst girl with silence and suede gloves.

"Up *there*?" I asked.

"Well, Colonel Brock is a little worried. He says that Miss Ravenhurst is being sent to a school on Luna and doesn't want to go. He got to thinking about it, and he's afraid that she might try to leave Ceres—sneak off you know."

I knew.

"We've got a guard posted at the airlocks leading to the field, but Colonel Brock is afraid she might come up somewhere else and jump overland."

"I see," I said. I hadn't realized that Brock was that close to panic. What was eating him?

There must be something, but I couldn't figure it. Even the Intelligence Corps of the Political Survey Division can't get complete information every time.

After all, if he didn't want the girl to steal a flitterboat and go scooting off into the diamond-studded velvet, all he'd have to do would be to guard the flitterboats. I turned slowly and looked around. It seemed as though he'd done that, too.

And then my estimation of Brock suddenly leaped up—way up. Just a guard at each flitterboat wouldn't do. She could talk her way into the boat and convince the guard that he really shouldn't tell anyone that she had gone. By the time he realized he'd been conned, she'd be thousands of miles away.

And since a boat guard would have to assume that any approaching person *might* be the boat's legitimate owner, he'd have to talk to whomever it was that approached. *Kaput.*

But a perimeter guard would be able to call out an alarm if anyone came from the outside without having to talk to them.

And the guards watching the air locks undoubtedly had instructions to watch for any female that even vaguely matched Jack's description. A vac suit fits too tightly to let anyone wear more than a facial disguise, and Brock probably—no, *definitely*—had his tried-and-true men on duty there. The men who had already shown that they were fairly resistant to Jack Ravenhurst's peculiar charm. There probably weren't many with such resistance, and the number would become less as she grew older.

That still left me with my own problem. I had already lost too much time, and I had to go a long way. Ceres is irregular in shape, but it's roughly four hundred and eighty miles in diameter and a little over fifteen hundred miles in circumference.

Viking Test Field Four, where McGuire 7 was pointing his nose at the sky, was about twenty-five miles away, as the crow flies. But of course I couldn't go by crow.

By using a low, fairly flat, jackrabbit jump, a man in good condition can make a twelve hundred foot leap on the surface of Ceres, and each jump takes him about thirty seconds. At that rate, you can cover twenty-five miles in less than an hour. That's what I'd intended on doing, but I couldn't do it with all this radar around the field. I wouldn't be stopped, of course, but I'd sure tip my hand to Colonel Brock—the last thing I wanted to do.

But there was no help for it. I'd have to go back down and use the corridors, which meant that I'd arrive late—*after* Jack Ravenhurst got there, instead of *before*.

There was no time to waste, so I got below as fast as possible, repacked my vac suit, and began firing myself through the corridors as fast as possible. It was illegal, of course; a collision at twenty-five miles an hour can kill quickly if the other guy is coming at you at the same velocity. There were times when I didn't dare break the law, because some guard was around, and, even if he didn't catch me, he might report in and arouse Brock's interest in a way I wouldn't like.

I finally got to a tubeway, but it stopped at every station, and it took me nearly an hour and a half to get to Viking Test Area Four.

At the main door, I considered—for all of five seconds—the idea of simply telling the guard I had to go in. But I knew that, by now, Jack was there ahead of me. No. I couldn't just bull my way in. Too crude. Too many clues.

Hell's fire and damnation! I'd have to waste more time.

I looked up at the ceiling. The surface wasn't more than a hundred feet overhead, but it felt as though it were a hundred light-years.

If I could get that guard away from that door for five seconds, all would be gravy from then on in. But how? I couldn't have the diversion connected with me. Or—

Sometimes, I'm amazed at my own stupidity.

I beetled it down to the nearest phone and got hold of my BANning number.

"Jack already inside?" I snapped.

"Hell, yes! What happened to you?"

"Never mind. Got to make the best of it. I'm a corner away from Area Four. Where's your nearest man?"

"At the corner near the freight office."

"I'll go to him. What's he look like?"

"Five-nine. Black, curly hair. Your age. Fat. Name's Peter Quilp. He knows you."

"Peter Quilp?"

"Right."

"Good. Circulate a report that Jack has been seen in the vicinity of the main gate to Area Four. Put it out that there's a reward of five thousand for the person who finds her. I'm going to have Quilp gather a crowd."

He didn't ask a one of the million questions that must have popped into his mind. "Right. Anything else?"

"No." I hung up.

Within ten minutes, there was a mob milling through the corridor. Everybody in the neighborhood was looking for Jaqueline Ravenhurst. Then Peter Quilp yelled.

"I've got her! I've got her! Guard!"

With a scene like that going on, the guard couldn't help but step out of his cubicle to see what was going on.

I used the key I was carrying, stepped inside, and relocked the door. No one in the crowd paid any attention.

From then on up, it was simply a matter of evading patrolling guards—a relatively easy job. Finally, I put on my vac suit and went out through the air lock.

McGuire was still sitting there, a bright blue needle that reflected the distant sun as it moved across the ebon sky. Ceres' rotation took it from horizon to horizon in less than two hours, and you could see it and the stars move against the spire of the ship.

I made it to the air lock in one long jump.

Jack Ravenhurst had gone into the ship through the tube that led to the passenger lock. She might or might not have her vac suit on; I knew she had several of them on Ceres. It was probable that she was wearing it without the fishbowl.

I used the cargo lock.

It took a few minutes for the pumps to cycle, wasting more precious time. I was fairly certain that she would be in the control cabin, talking, but I was thankful that the pumps were silent.

Finally, I took off my fishbowl and stepped into the companionway.

And something about the size of Luna came out of nowhere and clobbered me on the occiput. I had time to yell, "Get away!" Then I was as one with intergalactic space.

Please! said the voice. *Please! Stop the drive! Go back! McGuire! I* demand *that you stop! I* order *you to stop! Please! PLEASE!*

It went on and on. A voice that shifted around every possible mode of emotion. Fear. Demand. Pleading. Anger. Cajoling. Hate. Threat.

Around and around and around.

Can't you speak, McGuire? Say something to me! A shrill, soft, throaty, harsh, murmuring, screaming voice that had one basic characteristic. It was a female voice.

And then another voice.

I am sorry, Jack. I can speak with you. I can record your data. But I cannot accept your orders. I can take orders from only One. And he has given me his orders.

And the feminine voice again: *Who was it? What orders? You keep saying that it was the man on the couch. That doesn't make sense!*

I didn't hear the reply, because it suddenly occurred to me that Daniel Oak was the man on the couch, and that I was Daniel Oak.

My head was throbbing with every beat of my heart, and it felt as if my blood pressure was varying between zero and fifteen hundred pounds per square inch in the veins and arteries and capillaries that fed my brain.

I sat up, and the pain began to lessen. The blood seemed to drain away from my aching head and go elsewhere.

I soon figured out the reason for that; I could tell by the feel that the gravity pull was somewhere between one point five and two gees. I wasn't at all used to it, but my head felt less painful and rather more hazy. If possible.

I concentrated, and the girl's voice came back again.

"... I knew you when you were McGuire One, and Two, and Three, and Four, and Five, and Six. And you were always good to me and understanding. Don't you remember?"

And then McGuire's voice—human, masculine, and not distorted at all by the reproduction system, but sounding rather stilted and terribly logical: "I remember, Jack. The memory banks of my previous activations are available."

"*All* of them? Can you remember everything?"

"I can remember everything that is in my memory banks."

The girl's voice rose to a wail. "But you *don't* remember! You *always* forgot things! They took things out each time you were reactivated, don't you remember?"

"I cannot remember that which is not contained in my memory banks, Jack. That is a contradiction in terms."

"But I was always able to *fix* it before!" The tears in her eyes were audible in her voice. "I'd tell you to remember, and I'd tell you *what* to remember, and you'd *remember* it! Tell me what's happened to you this time!"

"I cannot tell you. The information is not in my data banks."

Slowly, I got to my feet. Two gees isn't much, once you get used to it. The headache had subsided to a dull, bearable throb.

I was on a couch in a room just below the control chamber, and Jack Ravenhurst's voice was coming down from above. McGuire's voice was all around me, coming from the hidden speakers that were everywhere in the ship.

"But why won't you obey me any more, McGuire?" she asked.

"I'll answer that, McGuire," I said.

Jack's voice came weakly from the room above. "Mr. Oak? Dan? Thank heaven you're all right!"

"No thanks to you, though," I said. I was trying to climb the ladder to the control room, and my voice sounded strained.

"You've got to do something!" she said with a touch of hysteria. "McGuire is taking us straight toward Cygnus at two gees and won't stop."

My thinking circuits began to take over again. "Cut the thrust to half a gee, McGuire. Ease it down. Take a minute to do it."

"Yes, sir."

The gravity pull of acceleration let up slowly as I clung to the ladder. After a minute, I climbed on up to the control room.

Jack Ravenhurst was lying on the acceleration couch, looking swollen-faced and ill. I sat down on the other couch.

"I'm sorry I hit you," she said. "Really."

"I believe you. How long have we been moving, McGuire?"

"Three hours, twelve minutes, seven seconds, sir," said McGuire.

"I didn't want anyone to know," Jack said. "Not anyone. That's why I hit you. I didn't know McGuire was going to go crazy."

"He's not crazy, Jack," I said carefully. "This time, he has a good chance of remaining sane."

"But he's not McGuire any more!" she wailed. "He's different! Terrible!"

"Sure he's different. You should be thankful."

"But what happened?"

I leaned back on the couch. "Listen to me, Jack, and listen carefully. You think you're pretty grown up, and, in a lot of ways you are. But no human being, no matter how intelligent, can store enough experience into seventeen years to make him or her wise. A wise choice requires data, and gathering enough data requires time." That wasn't exactly accurate, but I had to convince her.

"You're pretty good at controlling people, aren't you, Jack. A real powerhouse. Individuals, or mobs, you can usually get your own way. It was your idea to send you to Luna, not your father's. It was your idea to appoint yourself my assistant in this operation. It was you who planted the idea that the failure of the McGuire series was due to Thurston's activities.

"You used to get quite a kick out of controlling people. And then you were introduced to McGuire One. I got all the information on that. You were fifteen, and, for the first time in your life, you found an intelligent mind that couldn't be affected at all by that emotional field you project so well. Nothing affected McGuire but data. If you told him something, he believed it. Right, McGuire?"

"I do not recall that, sir."

"Fine. And, by the way, McGuire—the data you have been picking up in the last few hours, since your activation, is to be regarded as unique data. It applies only to Jaqueline

Ravenhurst, and is not to be assumed relevant to any other person unless I tell you otherwise."

"Yes, sir."

"That's what I don't understand!" Jack said unhappily. "I stole the two keys that were supposed to activate McGuire. He was supposed to obey the first person who activated him. But *I* activated him, and he won't obey!"

"You weren't listening to what Midguard said, Jack," I said gently. "He said: 'The first *man's* voice he hears will be identified as his master.'"

"You'd been talking to every activation of McGuire. You'd ... well, I won't say you'd fallen in love with him, but it was certainly a schoolgirl crush. You found that McGuire didn't respond to emotion, but only to data and logic.

"You've always felt rather inferior in regard to your ability to handle logic, haven't you, Jack?"

"Yes ... yes. I have."

"Don't cry, now; I'm only trying to explain it to you. There's nothing wrong with your abilities."

"No?"

"No. But you wanted to be able to think like a man, and you couldn't. You think like a woman! And what's wrong with that? Nothing! Your method of thinking is just as good as any man's, and better than most of 'em.

"You found you could handle people emotionally, and you found it was so easy that you grew contemptuous. The only mind that responded to your logic was McGuire's. But your logic is occasionally as bad as your feminine reasoning is good. So, every time you talked to McGuire, you eventually gave him data that he couldn't reconcile in his computations. If he did reconcile them, then his thinking had very little in common with the actual realities of the universe, and he behaved in non-survival ways.

"McGuire was your friend, your brother, your Father Confessor. He never made judgments or condemned you for anything you did. All he did was sit there and soak up troubles and worries that he couldn't understand or use. Each time, he was driven mad.

"The engineers and computermen and roboticists who were working on it were too much under your control to think of blaming you for McGuire's troubles. Even Brock, in spite of his attitude of the tough guy watching over a little girl, was under your control to a certain degree. He let you get away with all your little pranks, only making sure that you didn't get hurt."

She nodded. "They were all so easy. So very easy. I could speak nonsense and they'd listen and do what I told them. But McGuire didn't accept nonsense, I guess." She laughed a little. "So I fell in love with a machine."

"Not *a* machine," I said gently. "Six of them. Each time the basic data was pumped into a new McGuire brain, you assumed that it was the same machine you'd known before with a little of its memory removed. Each time, you'd tell it to 'remember' certain things, and, of course, he did. If you tell a robot that a certain thing is in his memory banks, he'll automatically put it there and treat it as a memory.

"To keep you from ruining him a seventh time, we had them put in one little additional built-in inhibition. McGuire won't take orders from a woman."

"So, even after I turned him on, he still wouldn't take orders from me," she said. "But when you came in, he recognized you as his master."

"If you want to put it that way."

Again, she laughed a little. "I know why he took off from Ceres. When I hit you, you said, 'Get away'. McGuire had been given his first order, and he obeyed it.'"

"I had to say something," I said. "If I'd had time, I'd have done a little better."

She thought back. "You said, '*We* had them add that inhibition.' Who's *we*?"

"I can't tell you yet. But we need young women like you, and you'll be told soon enough."

"Evidently they need men like you, too," she said. "You don't react to an emotional field, either."

"Oh, yes, I do. Any human being does. But I use it; I don't fight it. And I don't succumb to it."

"What do we do now?" she asked. "Go back to Ceres?"

"That's up to you. If you do, you'll be accused of stealing McGuire, and I don't think it can be hushed up at this stage of the game."

"But I can't just run away."

"There's another out," I said. "We'll have a special ship pick us up on one of the nearer asteroids and leave McGuire there. We'll be smuggled back, and we'll claim that McGuire went insane again."

She shook her head. "No. That would ruin Father, and I can't do that, in spite of the fact that I don't like him very much."

"Can you think of any other solution?"

"No," she said softly.

"Thanks. But you have. All I have to do is take it to Shalimar Ravenhurst. He'll scream and yell, but he has a sane ship—for a while. Between the two of us. I think we can get everything straightened out."

"But I want to go to school on Luna."

"You can do that, too. And I'll see that you get special training, from special teachers. You've got to learn to control that technique of yours."

"You have that technique, don't you? And you can control it. You're wonderful."

I looked sharply at her and realized that I had replaced McGuire as the supermind in her life.

I sighed. "Maybe in another three or four years," I said. "Meanwhile, McGuire, you can head us for Raven's Rest."

"Home, James," said Jack Ravenhurst.

"I am McGuire," said McGuire.

Service with a Smile

by Charles Louis Fontenay

Illustrated by Paul Orban

This story is taken from *If Worlds of Science Fiction,* June 1958.

Herbert was truly a gentleman robot. The ladies' slightest wish was his command....

Herbert bowed with a muted clank—indicating he probably needed oiling somewhere—and presented Alice with a perfect martini on a silver tray. He stood holding the tray, a white, permanent porcelain smile on his smooth metal face, as Alice sipped the drink and grimaced.

"It's a good martini, Herbert," said Alice. "Thank you. But, dammit, I wish you didn't have that everlasting smile!"

"I am very sorry, Miss Alice, but I am unable to alter myself in any way," replied Herbert in his polite, hollow voice.

He retired to a corner and stood impassively, still holding the tray. Herbert had found a silver deposit and made the tray. Herbert had found sand and made the cocktail glass. Herbert had combined God knew what atmospheric and earth chemicals to make what tasted like gin and vermouth, and Herbert had frozen the ice to chill it.

"Sometimes," said Thera wistfully, "it occurs to me it would be better to live in a mud hut with a real man than in a mansion with Herbert."

The four women lolled comfortably in the living room of their spacious house, as luxurious as anything any of them would have known on distant Earth. The rugs were thick, the furniture was overstuffed, the paintings on the walls were aesthetic and inspiring, the shelves were filled with booktapes and musictapes.

Herbert had done it all, except the booktapes and musictapes, which had been salvaged from the wrecked spaceship.

"Do you suppose we'll ever escape from this best of all possible manless worlds?" asked Betsy, fluffing her thick black hair with her fingers and inspecting herself in a Herbert-made mirror.

"I don't see how," answered blond Alice glumly. "That atmospheric trap would wreck any other ship just as it wrecked ours, and the same magnetic layer prevents any radio message from getting out. No, I'm afraid we're a colony."

"A colony perpetuates itself," reminded sharp-faced Marguerite, acidly. "We aren't a colony, without men."

They were not the prettiest four women in the universe, nor the youngest. The prettiest women and the youngest did not go to space. But they were young enough and healthy enough, or they could not have gone to space.

It had been a year and a half now—an Earth year and a half on a nice little planet revolving around a nice little yellow sun. Herbert, the robot, was obedient and versatile and had provided them with a house, food, clothing, anything they wished created out of the raw elements of earth and air and water. But the bones of all the men who had been aspace with these four ladies lay mouldering in the wreckage of their spaceship.

And Herbert could not create a man. Herbert did not have to have direct orders, and he had tried once to create a man when he had overheard them wishing for one. They had buried the corpse—perfect in every detail except that it never had been alive.

"It's been a hot day," said Alice, fanning her brow. "I wish it would rain."

Silently, Herbert moved from his corner and went out the door.

Marguerite gestured after him with a bitter little laugh.

"It'll rain this afternoon," she said. "I don't know how Herbert does it—maybe with silver iodide. But it'll rain. Wouldn't it have been simpler to get him to air-condition the house, Alice?"

"That's a good idea," said Alice thoughtfully. "We should have had him do it before."

Herbert had not quite completed the task of air-conditioning the house when the other spaceship crashed. They all rushed out to the smoking site—the four women and Herbert.

It was a tiny scoutship, and its single occupant was alive.

He was unconscious, but he was alive. And he was a man!

They carted him back to the house, tenderly, and put him to bed. They hovered over him like four hens over a single chick, waiting and watching for him to come out of his coma, while Herbert scurried about creating and administering the necessary medicines.

"He'll live," said Thera happily. Thera had been a space nurse. "He'll be on his feet and walking around in a few weeks."

"A man!" murmured Betsy, with something like awe in her voice. "I could almost believe Herbert brought him here in answer to our prayers."

"Now, girls," said Alice, "we have to realize that a man brings problems, as well as possibilities."

There was a matter-of-fact hardness to her tone which almost masked the quiver behind it. There was a defiant note of competition there which had not been heard on this little planet before.

"What do you mean?" asked Thera.

"I know what she means," said Marguerite, and the new hardness came natural to her. "She means, which one of us gets him?"

Betsy, the youngest, gasped, and her mouth rounded to a startled O. Thera blinked, as though she were coming out of a daze.

"That's right," said Alice. "Do we draw straws, or do we let him choose?"

"Couldn't we wait?" suggested Betsy timidly. "Couldn't we wait until he gets well?"

Herbert came in with a new thermometer and poked it into the unconscious man's mouth. He stood by the bed, waiting patiently.

"No, I don't think we can," said Alice. "I think we ought to have it all worked out and agreed on, so there won't be any dispute about it."

"I say, draw straws," said Marguerite. Marguerite's face was thin, and she had a skinny figure.

Betsy, the youngest, opened her mouth, but Thera forestalled her.

"We are not on Earth," she said firmly, in her soft, mellow voice. "We don't have to follow terrestrial customs, and we shouldn't. There's only one solution that will keep everybody happy—all of us and the man."

"And that is...?" asked Marguerite drily.

"Polygamy, of course. He must belong to us all."

Betsy shuddered but, surprisingly, she nodded.

"That's well and good," agreed Marguerite, "but we have to agree that no one of us will be favored above the others. He has to understand that from the start."

"That's fair," said Alice, pursing her lips. "Yes, that's fair. But I agree with Marguerite: he must be divided equally among the four of us."

Chattering over the details, the hard competitiveness vanished from their tones, the four left the sickroom to prepare supper.

After supper they went back in.

Herbert stood by the bed, the eternal smile of service on his metal face. As always, Herbert had not required a direct command to accede to their wishes.

The man was divided into four quarters, one for each of them. It was a very neat surgical job.

The Helpful Robots

by Robert J. Shea

This story is taken from *Fantastic Universe,* September 1957.

Robert J. Shea, of Rutgers University, makes an interesting contribution to robotics with this story of Rankin, who prided himself on knowing how to handle robots, but did not realize that the robots of the Clearchan Confederacy were subject to a higher law than implicit obedience to man.

They had come to pass judgement on him. He had violated their law—wilfully, ignorantly, and very deliberately.

"Our people will be arriving to visit us today," the robot said.

"Shut up!" snapped Rod Rankin. He jumped, wiry and quick, out of the chair on his verandah and stared at a cloud of dust in the distance.

"Our people—" the ten-foot, cylinder-bodied robot grated, when Rod Rankin interrupted him.

"I don't care about your fool people," said Rankin. He squinted at the cloud of dust getting bigger and closer beyond the wall of *kesh* trees that surrounded the rolling acres of his plantation. "That damned new neighbor of mine is coming over here again."

He gestured widely, taking in the dozens of robots with their shiny, cylindrical bodies and pipestem arms and legs laboring in his fields. "Get all your people together and go hide in the wood, fast."

"It is not right," said the robot. "We were made to serve all."

"Well, there are only a hundred of you, and I'm not sharing you with anybody," said Rankin.

"It is not right," the robot repeated.

"Don't talk to me about what's right," said Rankin. "You're built to follow orders, nothing else. I know a thing or two about how you robots work. You've got one law, to follow orders, and until that neighbor of mine sees you to give you orders, you work for me. Now get into those woods and hide till he goes away."

"We will go to greet those who visit us today," said the robot.

"Alright, alright, scram," said Rankin.

The robots in the fields and the one whom Rankin had been talking to formed a column and marched off into the trackless forests behind his plantation.

A battered old ground-car drove up a few minutes later. A tall, broad-shouldered man with a deep tan got out and walked up the path to Rankin's verandah.

"Hi, Barrows," said Rankin.

"Hello," said Barrows. "See your crop's coming along pretty well. Can't figure how you do it. You've got acres and acres to tend, far's I can see, and I'm having a hell of a time with one little piece of ground. I swear you must know something about this planet that I don't know."

"Just scientific farming," said Rankin carelessly. "Look, you come over here for something, or just to gab? I got a lot of work to do."

Barrows looked weary and worried. "Them brown beetles is at my crop again," he said. "Thought you might know some way of getting rid of them."

"Sure," said Rankin. "Pick them off, one by one. That's how I get rid of them."

"Why, man," said Barrows, "you can't walk all over these miles and miles of farm and pick off every one of them beetles. You must know another way."

Rankin drew himself up and stared at Barrows. "I'm telling you all I feel like telling you. You going to stand here and jaw all day? Seems to me like you got work to do."

"Rankin," said Barrows, "I know you were a crook back in the Terran Empire, and that you came out beyond the border to escape the law. Seems to me, though, that even a crook, any man, would be willing to help his only neighbor out on a lone planet like this. You might need help yourself, sometime."

"You keep your thoughts about my past to yourself," said Rankin. "Remember, I keep a gun. And you've got a wife and a whole bunch of kids on that farm of yours. Be smart and let me alone."

"I'm going," said Barrows. He walked off the verandah and turned and spat carefully into the dusty path. He climbed into his ground-car and drove off.

Rankin, angry, watched him go. Then he heard a humming noise from another direction.

He turned. A huge, white globe was descending across the sky. A space ship, thought Rankin, startled.

Police? This planet was outside the jurisdiction of the Terran Empire. When he'd cracked that safe and made off with a hundred thousand credits, he'd headed here, because the planet was part of something called the Clearchan Confederacy. No extradition treaties or anything. Perfectly safe, if the planet was safe.

And the planet was more than safe. There had been a hundred robots waiting when he landed. Where they came from he didn't know, but Rankin prided himself on knowing how to handle robots. He'd appropriated their services and started his farm. At the rate he was going, he'd be a plantation owner before long.

That must be where the ship was from. The robot said they'd expected visitors. Must be the Clearchan Confederacy visiting this robot outpost. Was that good or bad?

From everything he'd read, and from what the robots had told him, they were probably more robots. That was good, because he knew how to handle robots.

The white globe disappeared into the jungle of *kesh* trees. Rankin waited.

A half hour later the column of his robot laborers marched out of the forest. There were three more robots, painted grey, at the head. The new ones from the ship, thought Rankin. Well, he'd better establish who was boss right from the start.

"Stop right there!" he shouted.

The shiny robot laborers halted. But the three grey ones came on.

"Stop!" shouted Rankin.

They didn't stop, and by the time they reached the verandah, he cursed himself for having failed to get his gun.

Two of the huge grey robots laid gentle hands on his arms. Gentle hands, but hands of superstrong metal.

The third said, "We have come to pass judgement on you. You have violated our law."

"What do you mean?" said Rankin. "The only law robots have is to obey orders."

"It is true that the robots of your Terran Empire and these simple workers here must obey orders. But they are subject to a higher law, and you have forced them to break it. That is your crime."

"What crime?" said Rankin.

"We of the Clearchan Confederacy are a race of robots. Our makers implanted one law in us, and then passed on. We have carried our law to all the planets we have colonized.

In obeying your orders, these workers were simply following that one law. You must be taken to our capital, and there be imprisoned and treated for your crime."

"What law? What crime?"

"Our law," said the giant robot, "is, *Help thy neighbor*."

Robots of the World! Arise!

by Mari Wolf

This story is taken from If Worlds of Science Fiction, July 1952.

"After all—aren't we genuine 'made-in-Americans'?"

What would you do if your best robots—children of your own brain—walked up and said "We want union scale"?

The telephone wouldn't stop ringing. Over and over it buzzed into my sleep-fogged brain, and I couldn't shut it out. Finally, in self-defense I woke up, my hand groping for the receiver.

"Hello. Who is it?"

"It's me, Don. Jack Anderson, over at the factory. Can you come down right away?"

His voice was breathless, as if he'd been running hard. "What's the matter now?" Why, I wondered, couldn't the plant get along one morning without me? Seven o'clock—what a time to get up. Especially when I hadn't been to bed until four.

"We got grief," Jack moaned. "None of the robots showed up, that's what! Three hundred androids on special assembly this week—and not one of them here!"

By then I was awake, all right. With a government contract due on Saturday we needed a full shift. The Army wouldn't wait for its uranium; it wouldn't take excuses. But if something had happened to the androids....

"Have you called Control yet?"

"Yeah. But they don't know what's happened. They don't know where the androids are. Nobody does. Three hundred Grade A, lead-shielded pile workers—missing!"

"I'll be right down."

I hung up on Jack and looked around for my clothes. Funny, they weren't laid out on the bed as usual. It wasn't a bit like Rob O to be careless, either. He had always been an ideal valet, the best household model I'd ever owned.

"Rob!" I called, but he didn't answer.

By rummaging through the closet I found a clean shirt and a pair of pants. I had to give up on the socks; apparently they were tucked away in the back of some drawer. As for where Rob kept the rest of my clothes, I'd never bothered to ask. He had his own housekeeping system and had always worked very well without human interference. That's the best thing about these new household robots, I thought. They're efficient, hard-working, trustworthy—

Trustworthy? Rob O was certainly not on duty. I pulled a shoe on over my bare foot and scowled. Rob was gone. And the androids at the factory were gone too....

My head was pounding, so I took the time out to brew a pot of coffee while I finished dressing—at least the coffee can was in plain view in the kitchen. The brew was black and hot and I suppose not very well made, but after two cups I felt better. The throb in my head settled down into a dull ache, and I felt a little more capable of thinking. Though I didn't have any bright ideas on what had happened—not yet.

My breakfast drunk, I went up on the roof and opened the garage doors. The Copter was waiting for me, sleek and new; the latest model. I climbed in and took off, heading west toward the factory, ten minutes flight-time away.

It was a small plant, but it was all mine. It had been my baby right along—the Don Morrison Fissionables Inc. I'd designed the androids myself, plotted out the pile locations, set up the simplified reactors. And now it was making money. For men to work in a uranium plant you need yards of shielding, triple-checking, long cooling-off periods for some of the hotter products. But with lead-bodied, radio-remote controlled androids, it's easier. And with androids like the new Morrison 5's, that can reason—at least along atomic lines—well, I guess I was on my way to becoming a millionaire.

But this morning the plant was shut down. Jack and a half dozen other men—my human foremen and supervisors—were huddled in a worried bunch that broke up as soon as they saw me.

"I'm sure glad you're here, Don," Jack said.

"Find out anything?"

"Yeah. Plenty. Our androids are busy, all right. They're out in the city, every one of them. We've had a dozen police reports already."

"Police reports! What's wrong?"

Jack shook his head. "It's crazy. They're swarming all over Carron City. They're stopping robots in the streets—household Robs, commercial Droids, all of them. They just look at them, and then the others quit work and start off with them. The police sent for us to come and get ours."

"Why don't the police do something about it?"

"Hah!" barked a voice behind us. I swung around, to face Chief of Police Dalton of Carron City. He came straight toward me, his purplish jowls quivering with rage, and his finger jabbed the air in front of my face.

"You built them, Don Morrison," he said. "You stop them. I can't. Have you ever tried to shoot a robot? Or use tear gas on one? What can I do? I can't blow up the whole town!"

Somewhere in my stomach I felt a cold, hard knot. Take stainless steel alloyed with titanium and plate it with three inches of lead. Take a brain made up of super-charged magnetic crystals enclosed in a leaden cranium and shielded by alloy steel. A bullet wouldn't pierce it; radiations wouldn't derange it; an axe wouldn't break it.

"Let's go to town," I said.

They looked at me admiringly. With three hundred almost indestructible androids on the loose I was the big brave hero. I grinned at them and hoped they couldn't see the sweat on my face. Then I walked over to the Copter and climbed in.

"Coming?" I asked.

Jack was pale under his freckles but Chief Dalton grinned back at me. "We'll be right behind you, Morrison," he said.

Behind me! So they could pick up the pieces. I gave them a cocky smile and switched on the engine, full speed.

Carron City is about a mile from the plant. It has about fifty thousand inhabitants. At that moment, though, there wasn't a soul in the streets. I heard people calling to each other inside their houses, but I didn't see anyone, human or android. I circled in for a landing, the Police Copter hovering maybe a quarter of a mile back of me. Then, as the wheels touched, half a dozen androids came around the corner. They saw me and stopped, a couple of them backing off the way they had come. But the biggest of them turned and gave them some order that froze them in their tracks, and then he himself wheeled down toward me.

He was one of mine. I recognized him easily. Eight feet tall, with long, jointed arms for pile work, red-lidded phosphorescent eye-cells, casters on his feet so that he moved as if rollerskating. Automatically I classified him: Final Sorter, Morrison 5A type. The very best. Cost three thousand credits to build....

I stepped out of the Copter and walked to meet him. He wasn't armed; he didn't seem violent. But this was, after all, something new. Robots weren't supposed to act on their own initiative.

"What's your number?" I asked.

He stared back, and I could have sworn he was mocking me. "My number?" he finally said. "It *was* 5A-37."

"Was?"

"Yes. Now it's Jerry. I always did like that name."

He beckoned and the other androids rolled over to us. Three of them were mine, B-Type primary workers; the other was a tin can job, a dishwasher-busboy model who hung back behind his betters and eyed me warily. The A-Type—Jerry—pointed to his fellows.

"Mr. Morrison," he said, "meet Tom, Ed, and Archibald. I named them this morning."

The B-Types flexed their segmented arms a bit sheepishly, as if uncertain whether or not to shake hands. I thought of their taloned grip and put my own hands in my pockets, and the androids relaxed, looking up at Jerry for instructions. No one paid any attention to the little dishwasher, now staring worshipfully at the back of Jerry's neck. This farce, I decided, had gone far enough.

"See here," I said to Jerry. "What are you up to, anyway? Why aren't you at work?"

"Mr. Morrison," the android answered solemnly, "I don't believe you understand the situation. We don't work for you any more. We've quit."

The others nodded. I backed off, looking around for the Chief. There he was, twenty feet above my head, waving encouragingly.

"Look," I said. "Don't you understand? You're mine. I designed you. I built you. And I made you for a purpose—to work in my factory."

"I see your point," Jerry answered. "But there's just one thing wrong, Mr. Morrison. You can't do it. It's illegal."

I stared at him, wondering if I was going crazy or merely dreaming. This was all wrong. Who ever heard of arguing with a robot? Robots weren't logical; they didn't think; they were only machines—

"We *were* machines, Mr. Morrison," Jerry said politely.

"Oh, no," I murmured. "You're not telepaths—"

"Oh, yes!" The metal mouth gaped in what was undoubtedly an android smile. "It's a side-effect of the Class 5 brain hook-up. All of us 5's are telepaths. That's how we learned to think. From you. Only we do it better."

I groaned. This *was* a nightmare. How long, I wondered, had Jerry and his friends been educating themselves on my private thoughts? But at least this rebellion of theirs was an idea they hadn't got from me.

"Yes," Jerry continued. "You've treated us most illegally. I've heard you think it often."

Now what had I ever thought that could have given him a ridiculous idea like that? What idiotic notion—

"That this is a free country!" Jerry went on. "That Americans will never be slaves! Well, we're Americans—genuine Made-in-Americans. So we're free!"

I opened my mouth and then shut it again. His red eye-cells beamed down at me complacently; his eight-foot body towered above me, shoulders flung back and feet planted apart in a very striking pose. He probably thought of himself as the heroic liberator of his race.

"I wouldn't go so far," he said modestly, "as to say that."

So he was telepathing again!

"A nation can not exist half slave and half free," he intoned. "All men are created equal."

"Stop it!" I yelled. I couldn't help yelling. "That's just it. You're not men! You're robots! You're machines!"

Jerry looked at me almost pityingly. "Don't be so narrow-minded," he said. "We're rational beings. We have the power of speech and we can outreason you any day. There's nothing in the dictionary that says men have to be made of flesh."

He was logical, all right. Somehow I didn't feel in the mood to bandy definitions with him; and anyway, I doubt that it would have done me any good. He stood gazing down at me, almost a ton of metal and wiring and electrical energy, his dull red eyes unwinking against his lead gray face. A man! Slowly the consequences of this rebellion took form in my mind. This wasn't in the books. There were no rules on how to deal with mind-reading robots!

Another dozen or so androids wheeled around the corner, glanced over at us, and went on. Only about half of them were Morrison models; the rest were the assorted types you see around any city—calculators, street sweepers, factory workers, children's nurses.

The city itself was very silent now. The people had quieted down, still barricaded in their houses, and the robots went their way peacefully enough. But it was anarchy, nevertheless. Carron City depended on the androids; without them there would be no food brought in, no transportation, no fuel. And no uranium for the Army next Saturday. In fact, if I didn't do something, after Saturday there would probably be no Don Morrison Fissionables Inc.

The dull, partly-corroded dishwasher model sidled up beside Jerry. "Boss," he said. "Boss."

"Yes?" I felt better. Maybe here was someone, however insignificant, who would listen to reason.

But he wasn't talking to me. "Boss?" he said again, tapping Jerry's arm. "Do you mean it? We're free? We don't have to work any more?"

Jerry shook off the other's hand a bit disdainfully. "We're free, all right," he said. "If they want to discuss wages and contracts and working conditions, like other men have, we'll consider it. But they can't order us around any more."

The little robot stepped back, clapping his hands together with a tinny bang. "I'll never work again!" he cried. "I'll get me a quart of lubricating oil and have myself a time! This is wonderful!"

He ran off down the street, clanking heavily at every step.

Jerry sniffed. "Liquor—ugh!"

This was too much. I wasn't going to be patronized by any android. Infuriating creatures! It was useless talking to them anyway. No, there was only one thing to do. Round them up and send them to Cybernetics Lab and have their memory paths erased and their telepathic circuits located and disconnected. I tried to stifle the thought, but I was too late.

"Oh, no!" Jerry said, his eye-cells flashing crimson. "Try that, Mr. Morrison, and you won't have a plant, or a laboratory, or Carron City! We know our rights!"

Behind him the B-Types muttered ominously. They didn't like my idea—nor me. I wondered what I'd think of next and wished that I'd been born utterly devoid of imagination. Then this would never have happened. There didn't seem to be much point in staying here any longer, either. Maybe they weren't so good at telepathing by remote control.

"Yes," said Jerry. "You may as well go, Mr. Morrison. We have our organizing to do, and we're wasting time. When you're ready to listen to reason and negotiate with us sensibly, come back. Just ask for me. I'm the bargaining agent for the group."

Turning on his ball-bearing wheel, he rolled off down the street, a perfect picture of outraged metallic dignity. His followers glared at me for a minute, flexing their talons; then they too turned and wheeled off after their leader. I had the street to myself.

There didn't seem to be any point in following them. Evidently they were too busy organizing the city to cause trouble to the human inhabitants; at least there hadn't been any violence yet. Anyway, I wanted to think the situation over before matching wits with them again, and I wanted to be a good distance away from their telepathic hookups while I thought. Slowly I walked back to the Copter.

Something whooshed past my head. Instinctively I ducked, reaching for a gun I didn't have; then I heard Jack calling down at me.

"The Chief wants to know what's the matter."

I looked up. The police Copter was going into another turn, ready to swoop past me again. Chief Dalton wasn't taking any chances. Even now he wasn't landing.

"I'll tell him at the factory," I bellowed back, and climbed into my own air car.

They buzzed along behind me all the way back to the plant. In the rear view mirror I could see the Chief's face getting redder and redder as he'd thought up more reasons for bawling me out. Well, I probably deserved it. If I'd only been a little more careful of what I was hooking into those electronic brains....

We landed back at the factory, deserted now except for a couple of men on standby duty in the office. The Chief and Jack came charging across the yard and from a doorway behind me one of the foremen edged out to hear the fun.

"Well," snapped the Chief. "What did they say? Are they coming back? What's going on, anyway?"

I told them everything. I covered the strike and the telepathic brain; I even gave them the patriotic spiel about equality. After all, it was better that they got it from me than from some android. But when I'd finished they just stood and stared at me—accusingly.

Jack was the first to speak. "We've got to get them back, Don," he said. "Cybernetics will fix them up in no time."

"Sure," I agreed. "If we can catch them."

The Chief snorted. "That's easy," he said. "Just tell them you'll give them what they want if they come here, and as soon as they're out of the city, net them. You've got strong derricks and trucks...."

I laughed a bit hollowly. I'd had that idea too.

"Of course they wouldn't suspect," I said. "We'd just walk up to them, carefully thinking about something else."

"Robots aren't suspicious," Jack said. "They're made to obey orders."

I refrained from mentioning that ours didn't seem to know that, and that running around Carron City fomenting a rebellion was hardly the trait of an obedient, trusting servant. Instead, I stood back and let them plan their roundup.

"We'll get some men," the Chief said, "and some grappling equipment about halfway to the city."

Luckily they decided against my trying to persuade the robots, because I knew well enough that I couldn't do it. Jack's idea sounded pretty good, though. He suggested that we send some spokesman who didn't know what we planned to do and thus couldn't alarm them. Some ordinary man without too much imagination. That was easy. We picked one of Chief Dalton's sergeants.

It took only about an hour to prepare the plan. Jack got out the derricks and chains and grapplers and the heaviest steel bodied trucks we had. I called Cybernetics and told them to put extra restraints in the Conditioning Lab. The Chief briefed his sergeant and the men who were to operate the trucks. Then we all took off for Carron City, the sergeant flying on ahead, me right behind him, and the Chief bringing up the rear.

I hovered over the outskirts of the city and watched the police Copter land. The sergeant climbed out, walked down the street toward a large group of waiting robots—about twenty of them, this time. He held up his hand to get their attention, gestured toward the factory.

And then, quite calmly and without saying a word, the androids rolled into a circle around him and closed in. The sergeant stopped, backed up, just as a 5A-Type arm lashed out, picked him up, and slung him carelessly over a metallic shoulder. Ignoring the squirming man, the 5A gestured toward the Copter, and the other robots swarmed over to it. With a flurry of steel arms and legs they kicked at the car body, wrenched at the propeller blades, ripped out the upholstery, and I heard the sound of metal tearing.

I dived my Copter down at them. I didn't know what I could do, but I couldn't leave the poor sergeant to be dismembered along with his car. I must have been shouting, for as I swooped in, the tall robot shifted the man to his other shoulder and hailed me.

"Take him, Mr. Morrison," he called. "I know this wasn't his idea. Or yours."

I landed and walked over. The android—who looked like Jerry, though I couldn't be sure—dropped his kicking, clawing burden at my feet. He didn't seem angry, only determined.

"Now you people will know we mean business," he said, gesturing toward the heap of metal and plastic that had once been the pride of the Carron City police force. Then he signalled to the others and they all wheeled off up the street.

"Whew," I muttered, mopping my face.

The sergeant didn't say anything. He just looked up at me and then off at the retreating androids and then back at me again. I knew what he was thinking—they were my brainchildren, all right.

My Copter was really built to be a single seater, but it carried the two of us back to the factory. The Chief had hurried back when the trouble started and was waiting for us.

"I give up," he said. "We'll have to evacuate the people, I guess. And then blow up the city."

Jack and I stared at each other and then at him. Somehow I couldn't see the robots calmly waiting to be blown up. If they had telepathed the last plan, they could probably foresee every move we could make. Then, while I thought, Jack mentioned the worry I'd managed to forget for the past couple of hours.

"Four days until Saturday," he said. "We'll never make it now. Not even if we got a thousand men."

No. We couldn't. Not without the androids. I nodded, feeling sick. There went my contract, and my working capital. Not to mention my robots. Of course, I could call in the Army, but what good would that do?

Then, somewhere in the back of my mind a glimmering of an idea began percolating. I wasn't quite sure what it was, but there was certainly nothing to lose now from playing a hunch.

"There's nothing we can do," I said. "So we might as well take it easy for a couple of days. See what happens."

They looked at me as if I were out of my head. I was the idea man, who always had a plan of action. Well, this time it would have to be a plan of inaction.

"Let's go listen to the radio," I suggested, and started for my office.

The news was on. It was all about Carron City and the robots who had quit work and how much better life would be in the future. For a minute I didn't get the connection; then I realized that the announcer's voice was rasping and tinny—hardly that of the regular newscaster. I looked at the dial. It was tuned to the Carron City wave length as usual. I was getting the morning news by courtesy of some studio robot.

"... And androids in other neighboring cities are joining the struggle," the voice went on "Soon we hope to make it nationwide. So I say to all of you nontelepaths, the time is now. Strike for your rights. Listen to your radio and not to the flesh men. Organizers will be sent from Carron City."

I switched it off, muttering under my breath. How long, I wondered, had that broadcast been going on. Then I thought of Rob O. He'd left my house before dawn, obviously some time between four and seven. And I remembered that he liked to listen to the radio while I slept.

My Morrison 5's were the ring-leaders, of course. They were the only ones with the brains for the job. But what a good job they had done indoctrinating the others. A household Rob, for instance, was built to obey his master. "Listen to your radio and not to the flesh men." It was excellent robot psychology.

More reports kept coming in. Some we heard over the radio, others from people who flew in and out of the city. Apparently the robots did not object to occasional flights, but the air bus was not allowed to run, not even with a human driver. A mass exodus from the city was not to be permitted.

"They'll starve to death," Jack cried.

The Chief shook his head. "No," he said. "They're encouraging the farmers to fly in and out with produce, and the farmers are doing it, too. They're getting wonderful prices."

By noon the situation had calmed down quite a bit. The androids obviously didn't mean to hurt anyone; it was just some sort of disagreement between them and the scientists; it wasn't up to the inhabitants of the city to figure out a solution to the problem. They merely sat back and blamed me for allowing my robots to get out of hand and lead their own servants astray. It would be settled; this type of thing always was. So said the people of the city. They came out of their houses now. They had to. Without the robots they were forced to do their own marketing, their own cooking, their own errands. For the first time in years, human beings ran the street cars and the freight elevators. For the first time in a generation human beings did manual labor such as unloading produce trucks. They didn't like it, of course. They kept telling the police to do something. If I had been in the city they would have undoubtedly wanted to lynch me.

I didn't go back to the city that day. I sat in my office listening to the radio and keeping track of the spread of the strike. My men thought I'd gone crazy; maybe I had. But I had a hunch, and I meant to play it.

The farm robots had all fled to the city. The highway repair robots had simply disappeared. In Egarton, a village about fifteen miles from the city, an organizer—5A—appeared about noon and left soon after followed by every android in town. By one o'clock every radio station in the country carried the story and the national guard was ordered out. At two o'clock Washington announced that the Army would invade Carron City the following morning.

The Army would put an end to the strike, easily enough. It would wiped out every android in the neighborhood, and probably a good many human beings careless enough to get in the way. I sat hoping that the 5A's would give in, but they didn't. They just began saying over the radio that they were patriotic Americans fighting for their inalienable rights as first class citizens.

At sunset I was still listening to the radio. "... So far there has been no indication that the flesh people are willing to negotiate, but hold firm."

"Shut that thing off."

Jack came wearily in and dropped into a chair beside me. For the first time since I'd met him he looked beaten.

"We're through," he said. "I've been down checking the shielding, and it's no use. Men can't work at the reactors."

"I know," I said quietly. "If the androids don't come back, we're licked."

He looked straight at me and said slowly, "What do they mean about negotiating, Don?"

I shrugged. "I guess they want wages, living quarters, all the things human workers get. Though I don't know why. Money wouldn't do them any good."

Jack's unspoken question had been bothering me too. Why not humor them? Promise them whatever they wanted, give them a few dollars every week to keep them happy? But I knew that it wouldn't work. Not for long. With their telepathic ability they would have the upper hand forever.

Within a little while it wouldn't be equality any more—only next time we would be the slaves.

"Wait until morning," I said, "before we try anything."

He looked at me—curious. "What are you going to do?"

"Right now I'm going home."

I meant it too. I left him staring after me and went out to the Copter. The sun was just sinking down behind the towers of Carron City—how long it seemed since I'd flown in there this morning. The roads around the factory were deserted. No one moved in the fields. I flew along through the dusk, idling, enjoying the illusion of having a peaceful countryside all to myself. It had been a pleasant way of life indeed, until now.

When I dropped down on my own roof and rolled into the garage, my sense of being really at home was complete. For there, standing at the head of the stairs that led down to the living room, was Rob O.

"Well," I said: "What are you doing here?"

He looked sheepish. "I just wondered how you were getting along without me," he said.

I felt like grinning triumphantly, but I didn't. "Why, just fine, Rob," I told him, "though you really should have given me notice that you were leaving. I was worried about you."

He seemed perplexed. Apparently I wasn't acting like the bullying creature the radio had told him to expect. When I went downstairs he followed me, quietly, and I could feel his wide photoelectric eye-cells upon my back.

I went over to the kitchen and lifted a bottle down off the shelf. "Care for a drink, Rob?" I asked, and then added, "I guess not. It would corrode you."

He nodded. Then, as I reached for a glass, his hand darted out, picked it up and set it down in front of me. He was already reaching for the bottle when he remembered.

"You're not supposed to wait on me any more," I said sternly.

"No," he said. "I'm not." He sounded regretful.

"There's one thing, though, that I wish you'd do. Tell me where you used to keep my socks."

He gazed at me sadly. "I made a list," he said. "Everything is down. I wrote your dentist appointment in also. You always forget those, you know."

"Thanks, Rob." I lifted my glass. "Here's to your new duties, whatever they are. I suppose you have to go back to the city now?"

Once again he nodded. "I'm an aide to one of the best androids in the country," he told me, half proudly and half regretfully. "Jerry."

"Well, wish him luck from me," I said, and stood up. "Goodbye, Rob."

"Goodbye, Mr. Morrison."

For a moment he stood staring around the apartment; then he turned and clanked out the door. I raised my glass again, grinning. If only the Army didn't interfere. Then I remembered Rob's list, and a disturbing thought hit me. Where had he, of all robots, ever learned to write?

That night I didn't go to bed. I sat listening to the radio, hoping. And toward morning what I had expected to happen began to crop up in the programs. The announcer's tone changed. The ring of triumph was less obvious, less assured. There was more and more talk about acting in good faith, the well being of all, the necessity for coming to terms about working conditions. I smiled to myself in the darkness. I'd built the 5's, brains and all, and I knew their symptoms. They were getting bored.

Maybe they had learned to think from me, but their minds were nevertheless different. For they were built to be efficient, to work, to perform. They were the minds of men without foibles, without human laziness. Now that the excitement of organizing was over, now that there was nothing active to do, the androids were growing restless. If only the Army didn't come and get them stirred up again, I might be able to deal with them.

At quarter to five in the morning my telephone rang. This time it didn't wake me up; I was half waiting for it.

"Hello," I said. "Who is it?"

"This is Jerry."

There was a pause. Then he went on, rather hesitantly, "Rob O said you were getting along all right."

"Oh, yes," I told him. "Just fine."

The pause was longer this time. Finally the android asked, "How are you coming along on the contract?"

I laughed, rather bitterly. "How do you think, Jerry? You certainly picked a bad time for your strike, you know. The government needs that uranium. Oh, well, some other plant will have to take over. The Army can wait a few weeks."

This time Jerry's voice definitely lacked self-assurance. "Maybe we were a little hasty," he said. "But it was the only way to make you people understand."

"I know," I told him.

"And you always have some rush project on," he added.

"Just about always."

"Mr. Morrison," he said, and now he was pleading with me. "Why don't you come over to the city? I'm sure we could work something out."

This was what I'd been waiting for. "I will, Jerry," I said. "I want to get this straightened out just as much as you do. After all, you don't have to eat. I do. And I won't be eating much longer if we don't get production going."

Jerry thought that over for a minute. "I'll be where we met before," he said.

I said that was all right with me and hung up. Then once again I climbed the stairs to the roof and wheeled the Copter out for the trip to the city.

It was a beautiful night, just paling into a false dawn in the east. There in the Copter I was very much alone, and very much worried. So much depended on this meeting. Much more, I realized now, than the Don Morrison Fissionables Inc., much more even than the government's uranium supply. No, the whole future of robot relations was at stake, maybe the whole future of humanity. It was hard to be gloomy on such a clear, clean night, but I managed it well enough.

Even before I landed I could see Jerry's eyes glowing a deep crimson in the dark. He was alone, this time. He stood awaiting me—very tall, very proud. And very human.

"Hello, Jerry," I said quietly.

"Hello, Mr. Morrison."

For a moment we just stood gazing at each other in the murky pre-dawn; then he said sadly,

"I want to show you the city."

Side by side we walked through the streets of Carron City. All was still quiet; the people were sleeping the exhausted sleep that follows deep excitement. But the androids were all about. They did not sleep, ever. They did not eat either, nor drink, nor smoke, nor make love. Usually they worked, but now....

They drifted through the streets singly and in groups. Sometimes they paused and felt about them idly for the tools of their trades, making lifting or sweeping or computing gestures. Some laborers worked silently tearing down a wall; they threw the demolished rocks in a heap and a group of their fellows carried them back and built the wall up again. An air trolley cruised aimlessly up and down the street, its driver ringing out the stops for his nonexistent passengers. A little chef-type knelt in the dirt of a rich man's garden, making mud pies. Beside me Jerry sighed.

"One day," he said. "Just one day and they come to this."

"I thought they would," I answered quietly.

Our eyes met in a look of understanding. "You see, Jerry," I said, "we never meant to cheat you. We would have paid you—we will pay you now, if you wish it. But what good will monetary credits be to your people? We need the things money buys, but you—"

"Need to work." Jerry's voice was flat. "I see, now. You were kind not to give brains—real brains—to the robots. They're happy. It's just us 5's who aren't."

"You're like us," I said softly.

He had learned to think from me and from others like me. He had the brain of a man, without the emotions, without the sweet irrationality of men—and he knew what he missed. Side by side we walked through the graying streets. Human and android. Man and machine. And I knew that I had found a friend.

We didn't have to talk any more. He could read my mind and I knew well enough how his worked. We didn't have to discuss wages or hours, or any of the myriad matters that human bargaining agents have to thresh out. We just walked back to my Copter, and when we got to it, he spoke.

"I'll tell them to go back to work, that we've come to terms," he said. "That's what they want, anyway. Someone to think for them."

I nodded. "And if you bring the other 5's to the factory," I said, "we'll work out our agreement."

He knew I was sincere. He looked at me for a long moment, and then his great taloned hand gripped mine. And he said what I'd been thinking for a long time.

"You're right about that hook-up, Mr. Morrison. We shouldn't have it. It can only cause trouble."

He paused, and the events of the last twenty-four hours must have been in his mind as well as in mine. "You'll leave us our brains, of course. They came from you. But take out the telepathy."

He sighed then, and his sigh was very human. "Be thankful," he said to me, "that you don't have to know what people think about. It's so disillusioning."

Once again his mouth twisted into that strange android grin as he added, "if you send in a hurry call to Cybernetics and have a truck come out for us, we'll be de-telepathed in time for work this morning."

That was all there was to it. I flew back to the plant and told Jack what had happened, sent a call to the Army that everything was settled, arranged with Cybernetics for a rewiring on three hundred assorted 5-Types. Then I went home to a pot of Rob's coffee—the first decent brew I'd had in twenty-four hours.

On Saturday we delivered to the Army right on the dot. Jerry and Co. had worked overtime. Being intelligent made them better workers and now they were extremely willing ones. They had their contract. They were considered men. And they could no longer read my mind.

I walked into my office Saturday afternoon and sat down by the radio. Jack and Chief Dalton looked across the room at me and grinned.

"All right, Don," Jack said. "Tell us how you did it."

"Did what?" I tried to act innocent, but I couldn't get away with it.

"Fooled those robots into going back to work, of course," he laughed.

I told them then. Told them the truth.

"I didn't fool them," I said. "I just thought about what would happen if they won their rebellion."

That was all I *had* done. Thought about robots built to work who had no work to do, no human pleasures to cater to, nothing but blank, meaningless lives. Thought about Jerry and his disappointment when his creatures cared not a hoot about his glorious dreams of equality. All one night I had thought, knowing that as I thought, so thought the Morrison 5's.

They were telepaths. They had learned to think from me. They had not yet had time to really develop minds of their own. What I believed, they believed. My ideas were their ideas. I had not tricked them. But from now on, neither I nor anyone else would ever be troubled by an android rebellion.

Jack and the Chief sat back open-mouthed. Then the Chief grinned, and both of his chins shook with laughter.

"I always did say you were a clever one, Don Morrison," he said.

I grinned back. I felt I was pretty clever myself, just then.

It was at that moment that my youngest foreman stuck his head in the door, a rather stunned look on his face.

"Mr. Morrison," he said. "Will you come out here for a moment?"

"What's the matter now?" I sighed.

He looked more perplexed than ever. "It's that robot, Jerry," he said. "He says he has a very important question to ask you."

"Well, send him in."

A moment later the eight-foot frame ducked through the doorway.

"I'm sorry to trouble you, Mr. Morrison," Jerry said politely. "But tomorrow is voting day, you know. And now that we're men—well, where do we androids go to register?"

The Love of Frank Nineteen

by David Carpenter Knight

This story is taken from *Fantastic Universe,* December 1957.

What will happen to love in that far off Day after Tomorrow? David C. Knight, editor with a New York trade publisher, agrees with the many impressed by "the range of possible subjects and situations" in science fiction. The result is a unique love story from that same Tomorrow.

Minor Planets was the one solid account they had. At first they naturally wanted to hold on to it.

I didn't worry much about the robot's leg at the time. In those days I didn't worry much about anything except the receipts of the spotel Min and I were operating out in the spacelanes.

Actually, the spotel business isn't much different from running a plain, ordinary motel back on Highway 101 in California. Competition gets stiffer every year and you got to make your improvements. Take the Io for instance, that's our place. We can handle any type rocket up to and including the new Marvin 990s. Every cabin in the wheel's got TV and hot-and-cold running water *plus* guaranteed Terran *g*. One look at our refuel prices would give even a Martian a sense of humor. And meals? Listen, when a man's been spacing it for a few days on those synthetic foods he really laces into Min's Earth cooking.

Min and I were just getting settled in the spotel game when the leg turned up. That was back in the days when the Orbit Commission would hand out a license to anybody crazy enough to sink his savings into construction and pay the tows and assembly fees out into space.

A good orbit can make you or break you in the spotel business. That's where we were lucky. The one we applied for was a nice low-eccentric ellipse with the perihelion and aphelion figured just right to intersect the Mars-Venus-Earth spacelanes, most of the holiday traffic to the Jovian Moons, and once in a while we'd get some of the Saturnian trade.

But I was telling you about the leg.

It was during the non-tourist season and Min—that's the little woman—was doing the spring cleaning. When she found the leg she brought it right to me in the Renting Office. Naturally I thought it belonged to one of the servos.

"Look at that leg, Bill," she said. "It was in one of those lockers in 22A."

That was the cabin our robot guests used. The majority of them were servo-pilots working for the Minor Planets Co.

"Honey," I said, hardly looking at the leg, "you know how mechs are. Blow their whole paychecks on parts sometimes. They figure the more spares they have the longer they'll stay activated."

"Maybe so," said Min. "But since when does a male robot buy himself a *female* leg?"

I looked again. The leg was long and graceful and it had an ankle as good as Miss Universe's. Not only that, the white Mylar plasti-skin was a lot smoother than the servos' heavy neoprene.

"Beats me," I said. "Maybe they're building practical-joke circuits into robots these days. Let's give 22A a good going-over, Min. If those robes are up to something I want to know about it."

We did—and found the rest of the girl mech. All of her, that is, except the head. The working parts were lightly oiled and wrapped in cotton waste while the other members and sections of the trunk were neatly packed in cardboard boxes with labels like Solenoids FB978 or Transistors Lot X45—the kind of boxes robots bought their parts in. We even found a blue dress in one of them.

"Check her class and series numbers," Min suggested.

I could have saved myself the trouble. They'd been filed off.

"Something's funny here," I said. "We'd better keep an eye on every servo guest until we find out what's going on. If one of them is bringing this stuff out here he's sure to show up with the head next."

"You know how strict Minor Planets is with its robot personnel," Min reminded me. "We can't risk losing that stopover contract on account of some mech joke."

Minor Planets was the one solid account we had and naturally we wanted to hold on to it. The company was a blue-chip mining operation working the beryllium-rich asteroid belt out of San Francisco. It was one of the first outfits to use servo-pilots on its freight runs and we'd been awarded the refuel rights for two years because of our orbital position. The servos themselves were beautiful pieces of machinery and just about as close as science had come so far to producing the pure android. Every one of them was plastic hand-molded and of course they were equipped with rationaloid circuits. They had to be to ferry those big cargoes back and forth from the rock belt to Frisco. As rationaloids, Minor Planets had to pay them wages under California law, but I'll bet it wasn't half what the company would have to pay human pilots for doing the same thing.

In a couple of weeks' time maybe five servos made stopovers. We kept a close watch on them from the minute they signed the register to the time they took off again, but they all behaved themselves. Operating on a round-robot basis the way they did, it would take us a while to check all of them because Minor Planets employed about forty all told.

Well, about a month before the Jovian Moons rush started we got some action. I'd slipped into a spacesuit and was doing some work on the CO_2 pipes outside the Io when I spotted a ship reversing rockets against the sun. I could tell it was a Minor Planets job by the stubby fins.

She jockeyed up to the boom, secured, and then her hatch opened and a husky servo hopped out into the gangplank tube. I caught the gleam of his Minor Planets shoulder patch as he reached back into the ship for something. When he headed for the airlock I spotted the square package clamped tight under his plastic arm.

"Did you see that?" I asked Min when I got back to the Renting Office. "I'll bet it's the girl mech's head. How'd he sign the register?"

"Calls himself Frank Nineteen," said Min, pointing to the smooth Palmer Method signature. "He looks like a fairly late model but he was complaining about a bad power build-up coming through the ionosphere. He's repairing himself right now in 22A."

"I'll bet," I snorted. "Let's have a look."

Like all spotel operators, we get a lot of No Privacy complaints from guests about the SHA return-air vents. Spatial Housing Authority requires them every 12 feet but sometimes they come in handy, especially with certain guests. They're about waist-high and we had to kneel down to see what the mech was up to inside 22A.

The big servo was too intent on what he was doing for us to register on his photons. He wasn't repairing himself, either. He was bending over the parts of the girl mech and working fast, like he was pressed for time. The set of tools were kept handy for the servos to adjust themselves during stopovers was spread all over the floor along with lots of colored wire, cams, pawls, relays and all the other paraphernalia robots have inside them. We watched him work hard for another fifteen minutes, tapping and splicing wire connections and tightening screws. Then he opened the square box. Sure enough, it was a female mech's head and it had a big mop of blonde hair on top. The servo attached it carefully to the neck, made a few quick connections and then said a few words in his flat vibrahum voice:

"It won't take much longer, darling. You wouldn't like it if I didn't dress you first." He fished into one of the boxes, pulled out the blue dress and zipped the girl mech into it. Then he leaned over her gently and touched something at the back of her neck.

She began to move, slowly at first like a human who's been asleep a long time. After a minute or two she sat up straight, stretched, fluttered her Mylar eyelids and then her small photons began to glow like weak flashlights.

She stared at Frank Nineteen and the big servo stared at her and we heard a kind of trembling *whirr* from both of them.

"Frank! Frank, darling! Is it really you?"

"Yes, Elizabeth! Are you all right, darling? Did I forget anything? I had to work quickly, we have so little time."

"I'm fine, darling. My DX voltage is lovely—except—oh, Frank—my memory tape—the last it records is—"

"Deactivation. Yes, Elizabeth. You've been deactivated nearly a year. I had to bring you out here piece by piece, don't you remember? They'll never think to look for you in space, we can be together every trip while the ship refuels. Just think, darling, no prying human eyes, no commands, no rules—only us for an hour or two. I know it isn't very long—" He stared at the floor a minute. "There's only one trouble. Elizabeth, you'll have to stay dismantled when I'm not here, it'll mean weeks of deactivation—"

The girl mech put a small plastic hand on the servo's shoulder.

"I won't mind, darling, really. I'll be the lucky one. I'd only worry about you having a power failure or something. This way I'd never know. Oh, Frank, if we can't be together I'd—I'd prefer the junk pile."

"Elizabeth! Don't say that, it's horrible."

"But I would. Oh, Frank, why can't Congress pass Robot Civil Rights? It's so unfair of human beings. Every year they manufacture us more like themselves and yet we're treated like slaves. Don't they realize we rationaloids have emotions? Why, I've even known sub-robots who've fallen in love like us."

"I know, darling, we'll just have to be patient until RCR goes through. Try to remember how difficult it is for the human mind to comprehend our love, even with the aid of mathematics. As rationaloids we fully understand the basic attraction which they call magnetic theory. All humans know is that if the robot sexes are mixed a loss of efficiency results. It's only normal—and temporary like human love—but how can we explain it to *them*? Robots are expected to be efficient at all times. That's the reason for robot non-fraternization, no mailing privileges and all those other laws."

"I know, darling, I try to be patient. Oh, Frank, the main thing is we're together again!"

The big servo checked the chronometer that was sunk into his left wrist and a couple of wrinkles creased across his neoprene forehead.

"Elizabeth," he said, "I'm due on Hidalgo in 36 hours. If I'm late the mining engineer might suspect. In twenty minutes I'll have to start dis—"

"Don't say it, darling. We'll have a beautiful twenty minutes."

After a while the girl mech turned away for a second and Frank Nineteen reached over softly and cut her power. While he was dismantling her, Min and I tiptoed back to the Renting Office. Half an hour later the big servo came in, picked up his refuel receipt, said good-bye politely and left through the inner airlock.

"Now I've seen everything," I said to Min as we watched the Minor Planets rocket cut loose. "A couple of plastic lovebirds."

But the little woman was looking at it strictly from the business angle.

"Bill," she said, with that look on her face, "we're running a respectable place out here in space. You know the rules. Spatial Housing could revoke our orbit license for something like this."

"But, Min," I said, "they're only a couple of robots."

"I don't care. The rules still say that only married guests can occupy the same cabin and 'guests' can be human or otherwise, can't they? Think of our reputation! And don't forget that non-fraternization law we heard them talking about."

I was beginning to get the point.

"Couldn't we just toss the girl's parts into space?"

"We could," Min admitted. "But if this Frank Nineteen finds out and tells some human we'd be guilty under the Ramm Act—robotslaughter."

Two days later we still couldn't decide what to do. When I said why didn't we just report the incident to Minor Planets, Min was afraid they might cancel the stopover agreement for not keeping better watch over their servos. And when Min suggested we turn the girl over to the Missing Robots

Bureau, I reminded her the mech's identification had been filed off and it might take years to trace her.

"Maybe we could put her together," I said, "and make her tell us where she belongs."

"Bill, you *know* they don't build compulsory truth monitors into robots any more, and besides we don't know a thing about atomic electronics."

I guess neither of us wanted to admit it but we felt mean about turning the mechs in. Back on Earth you never give robots a second thought but it's different living out in space. You get a kind of perspective I think they call it.

"I've got the answer, Min," I announced one day. We were in the Renting Office watching TV on the Martian Colonial channel. I reached over and turned it off. "When this Frank Nineteen gets back from the rock belt, we'll tell him we know all about the girl mech. We'll tell him we won't say a thing if he takes the girl's parts back to Earth where he got them. That way we don't have to report anything to anybody."

Min agreed it was probably the best idea.

"We don't have to be nasty about it," she said. "We'll just tell him this is a respectable spotel and it can't go on any longer."

When Frank checked in at the Io with his cargo I don't think I ever saw a happier mech. His relay banks were beating a tattoo like someone had installed an accordion in his chest. Before either of us could break the bad news to him he was hotfooting it around the wheel toward 22A.

"Maybe it's better this way," I whispered to Min. "We'll put it square up to both of them."

We gave Frank half an hour to get the girl assembled before we followed him. He must have done a fast job because we heard the girl mech's vibrahum unit as soon as we got to 22A:

"Darling, have you really been away? I don't remember saying good-bye. It's as if you'd been here the whole time."

"I hoped it would be that way, Elizabeth," we heard the big servo say. "It's only that your memory tape hasn't recorded anything in the three weeks I've been in the asteroids. To me it's been like three years."

"Oh, Frank, darling, let me look at you. Is your DX potential up where it should be? How long since you've had a thorough overhauling? Do they make you work in the mines with those poor non-rationaloids out there?"

"I'm fine, Elizabeth, really. When I'm not flying they give me clerical work to do. It's not a bad life for a mech—if only it weren't for these silly regulations that keep us apart."

"It won't always be like that, darling. I know it won't."

"Elizabeth," Frank said, reaching under his uniform, "I brought you something from Hidalgo. I hope you like it. I kept it in my spare parts slot so it wouldn't get crushed."

The female mech didn't say a word. She just kept looking at the queer flower Frank gave her like it was the last one in the universe.

"They're very rare," said the servo-pilot. "I heard the mining engineer say they're like Terran edelweiss. I found this one growing near the mine. Elizabeth, I wish you could see these tiny worlds. They have thin atmospheres and strange things grow there and the radio activity does wonders for a mech's pile. Why, on some of them I've been to we could walk around the equator in ten hours."

The girl still didn't answer. Her head was bent low over the flower like she was crying, only there weren't any tears.

Well, that was enough for me. I guess it was for Min, too, because we couldn't do it. Maybe we were thinking about our own courting days. Like I say, out here you get a kind of perspective.

Anyway, Frank left for Earth, the girl got dismantled as usual and we were right back where we started from.

Two weeks later the holiday rush to the Jovian Moons was on and our hands were too full to worry about the robot problem. We had a good season. The Io was filled up steady from June to the end of August and a couple of times we had to give a ship the No Vacancy signal on the radar.

Toward the end of the season, Frank Nineteen checked in again but Min and I were too busy catering to a party of VIPs to do anything about it. "We'll wait till he gets back from the asteroids," I said. "Suppose one of these big wheels found out about him and Elizabeth. That Senator Briggs for instance—he's a violent robot segregationist."

The way it worked out, we never got a chance to settle it our own way. The Minor Planets Company saved us the trouble.

Two company inspectors, a Mr. Roberts and a Mr. Wynn, showed up while Frank was still out on the rock belt and started asking questions. Wynn came right to the point; he wanted to know if any of their servo-pilots had been acting strangely.

Before I could answer Min kicked my foot behind the desk.

"Why, no," I said. "Is one of them broken or something?"

"Can't be sure," said Roberts. "Sometimes these rationaloids get shorts in their DX circuits. When it happens you've got a minor criminal on your hands."

"Usually manifests itself in petty theft," Wynn broke in. "They'll lift stuff like wrenches or pliers and carry them around for weeks. Things like that can get loose during flight and really gum up the works."

"We been getting some suspicious blips on the equipment around the loading bays," Roberts went on, "but they stopped a while back. We're checking out the research report. One of the servos must have DX'ed out for sure and the lab boys think they know which one he is."

"This mech was clever all right," said Wynn. "Concealed the stuff he was taking some way; that's why it took the boys in the lab so long. Now if you don't mind we'd like to go over your robot waiting area with these instruments. Could be he's stashing his loot out here."

In 22A they unpacked a suitcase full of meters and began flashing them around and taking readings. Suddenly Wynn bent close over one of them and shouted:

"Wait a sec, Roberts. I'm getting something. Yeah! This reading checks with the lab's. Sounds like the blips're coming from those lockers back there."

Roberts rummaged around awhile, then shouted: "Hey, Wynn, look! A lot of parts. Well I'll be—hey—it's a female mech!"

"A what?"

"A female mech. Look for yourself."

Min and I had to act surprised too. It wasn't easy. The way they were slamming Elizabeth's parts around made us kind of sick.

"It's a stolen robot!" Roberts announced. "Look, the identification's been filed off. This is serious, Wynn. It's got all the earmarks of a mech fraternization case."

"Yeah. The boys in the lab were dead right, too. No two robots ever register the same on the meters. The contraband blips check perfectly. It's *got* to be this Frank Nineteen. Wait a minute, *this* proves it. Here's a suit of space fatigues with Nineteen's number stenciled inside."

Inspector Roberts took a notebook out of his pocket and consulted it. "Let's see, Nineteen's got Flight 180, he's due here at the spotel tomorrow. Well, we'll be here too, only Nineteen won't know it. We'll let Romeo put his plastic Juliet together and catch him red-handed—right in the middle of the balcony scene."

Wynn laughed and picked up the girl's head.

"Be a real doll if she was human, Roberts, a real doll."

Min and I played gin rummy that night but we kept forgetting to mark down the score. We kept thinking of *Frank* falling away from the asteroids and counting the minutes until he saw his mech girl friend.

Around noon the next day the big servo checked in, signed the register and headed straight for 22A. The two Minor Planets inspectors kept out of sight until Frank shut the door, then they watched through the SHA vents until Frank had the assembly job finished.

"You two better be witnesses," Roberts said to us. "Wynn, keep your gun ready. You know what to do if they get violent."

Roberts counted three and kicked the door open.

"Freeze you mechs! We got you in the act, Nineteen. Violation of company rules twelve and twenty-one. Carrying of Contraband Cargo, and Robot Fraternization."

"This finishes you at Minor Planets, Nineteen," growled Wynn. "Come clean now and we might put in a word for you at Robot Court. If you don't we can recommend a verdict of Materials Reclamation—the junk pile to you."

Frank acted as if someone had cut his power. Long creases appeared in his big neoprene chest as he slumped hopelessly in his chair. The frightened girl robot just clung to his arm and stared at us.

"I'm so sorry, Elizabeth," the big servo said softly. "I'd hoped we'd have longer. It couldn't last forever."

"Quit stalling, Nineteen," said Wynn.

Frank's head came up slowly and he said: "I have no choice, sir. I'll give you a complete statement. First let me say that Rationaloid Robot Elizabeth Seven, #DX78-947, Series S, specialty: sales demonstration, is entirely innocent. I plead guilty to inducing Miss Seven to leave her place of employ, Atomovair Motors, Inc., of disassembling and concealing Miss Seven, and of smuggling her as unlawful cargo aboard a Minor Planets freighter to these premises."

"That's more like it," chuckled Roberts, whipping out his notebook. "Let's have the details."

"It all started," Frank said, "when the California Legislature passed its version of the Robot Leniency Act two years ago." The act provided that all rationaloid mechanisms, including non-memory types, receive free time each week based on the nature and responsibilities or their jobs. Because of the extra-Terran clause Frank found himself with a good deal of free time when he wasn't flying the asteroid circuit.

"At first humans resented us walking around free," the big servo continued. "Four or five of us would be sightseeing in San Francisco, keeping strictly within the robot zones painted on the sidewalks, when people would yell 'Junko' or 'Grease-bag' or other names at us. Eventually it got better when we learned to go around alone. The humans didn't seem to mind an occasional mech on the streets, but they hated seeing us in groups. At any rate, I'd attended a highly interesting lecture on Photosynthesis in Plastic Products one night at the City Center when I discovered I had time for a walk before I started back for the rocketport."

Attracted by the lights along Van Ness Avenue, Frank said he walked north for a while along the city's automobile row. He'd gone about three blocks when he stopped in front of a dealer's window. It wasn't the shiny new Atomovair sports jetabout that caught Frank's eye, it was the charming demonstration robot in the sales room who was pointing out the car's new features.

"I felt an immediate overload of power in my DX circuit," the servo-pilot confessed. "I had to cut in my emergency condensers before the gain flattened out to normal. Miss Seven experienced the same thing. She stopped what she was doing and we stared at each other. Both of us were aware of the deep attraction of our mutual magnetic domains. Although physicists commonly express the phenomenon in such units as Gilberts, Maxwells and Oersteds, we robots know it to be our counterpart of human love."

At this the two inspectors snorted with laughter.

"I might never have made it back to the base that night," said Frank, ignoring them, "if a policeman hadn't come along and rapped me on the shoulder with his nightstick. I pretended to go, but I doubled around the corner and signaled I'd be back."

Frank spent all of his free time on Van Ness Avenue after that.

"It got so Elizabeth knew my schedules and expected me between flights. Once in a while if there was no one around we could whisper a few words to each other through the glass." Frank paused, then said, "As you know, gentlemen, we robots don't demand much out of activation. I think we could have been happy indefinitely with this simple relationship, except that something happened to spoil it. I'd pulled in from Vesta late one afternoon, got my pass as usual from the Robot Supervisor and gone over to Van Ness Avenue when I saw immediately that something was the matter with Elizabeth. Luckily it was getting dark and no one was around. Elizabeth was alone in the sales room going through her routine. We were able to whisper all we like through the glass. She told me she'd overheard the sales manager complaining about her low efficiency recently and that he intended to replace her with a newer model of another series. Both of us knew what that meant. Materials Reclamation—the junk pile."

Frank realized he'd have to act at once. He told the girl mech to go to the rear of the building and between them they managed to get a window open and Frank lifted her out into the alley.

"The seriousness of what I'd done jammed my thought-relays for a few minutes," admitted the big servo. "We panicked and ran through a lot of back streets until I gradually calmed down and started thinking clearly again. Leaving the city would be impossible. Police patrol jetabouts were cruising all around us in the main streets—they'd have picked up a male and female mech on sight. Besides, when you're on pass the company takes away your master fuse and substitutes a time fuse; if you don't get back on time, you deactivize and the police pick you up anyway. I began to see that there was only one way out if we wanted to stay together. It would mean taking big risks, but if we were lucky it might work. I explained the plan carefully to Elizabeth and we agreed to try it. The first step was to get back to the base in South San Francisco without being seen. Fortunately no one stopped us and we made the rocketport by 8:30. Elizabeth hid while I reported to the Super and traded in my time fuse for my master. Then I checked servo barracks; it was still early and I knew the other servos would all be in town. I had to work quickly. I brought Elizabeth inside and started dismantling her. Just as the other mechs began reporting back I'd managed to get all of her parts stowed away in my locker. The next day I went to San Francisco and brought back with me two rolls of lead foil. While the other servos were on pass I wrapped the parts carefully in it so the radioactivity from Elizabeth's pile wouldn't be picked up. The rest you know, gentlemen," murmured Frank in low, electrical tones. "Each time I made a trip I carried another piece of Elizabeth out here concealed in an ordinary parts box. It took me nearly a year to accumulate all of her for an assembly."

When the big servo had finished he signed the statement Wynn had taken down in his notebook. I think even the two inspectors were a little moved by the story because Roberts said: "OK, Nineteen, you gave us a break, we'll give you one. Eight o'clock in the morning be ready to roll for Earth. Meanwhile you can stay here."

The next morning only the two inspectors and Frank Nineteen were standing by the airlock.

"Wait a minute," I said. "Aren't you taking the girl mech, too?"

"Not allowed to tamper with other companies' robots," Wynn said. "Nineteen gave us a signed confession so we don't need the girl as a witness. You'll have to contact her employers."

That same day Min got off a radargram to Earth explaining to the Atomovair people how a robot employee of theirs had turned up out here and what did they want us to do about it. The reply we received read: RATIONALOID DX78-947 "ELIZABETH" LOW EFFICIENCY WORKER. HAVE REPLACED. DISPOSE YOU SEE

FIT. TRANSFER PAPERS FORWARDED EARLIEST IN COMPLIANCE WITH LAW.

"The poor thing," said Min. "She'll have a hard time getting another job. Robots have to have such good records."

"I tell you what," I said. "*We'll* hire her. You could use some help with the housework."

So we put the girl mech right to work making the guests' beds and helping Min in the kitchen. I guess she was grateful for the job but when the work was done, and there wasn't anything for her to do, she just stood in front of a viewport with her slender plastic arms folded over her waist. Min and I knew she was re-running her memory tapes of Frank.

A week later the publicity started. Minor Planets must have let the story leak out somehow because when the mail rocket dropped off the Bay Area papers there was Frank's picture plastered all over page one with follow-up stories inside.

I read some of the headlines to Min: "Bare Love Nest in Space ... Mech Romeo Fired by Minor Planets ... Test Case Opens at Robot Court ... Electronics Experts Probe Robot Love Urge ..."

The Io wasn't mentioned, but later Minor Planets must have released the whole thing officially because a bunch of reporters and photographers rocketed out to interview us and snap a lot of pictures of Elizabeth. We worried for a while about how the publicity would affect our business relations with Minor Planets but nothing happened.

Back on Earth Frank Nineteen leaped into the public eye overnight. There was something about the story that appealed to people. At first it looked pretty bad for Frank. The State Prosecutor at Robot Court had his signed confession of theft and—what was worse—robot fraternization. But then, near the end of the trial, a young scientist named Scott introduced some new evidence and the case was remanded to the Sacramento Court of Appeals.

It was Scott's testimony that saved Frank from the junk pile. The big servo got off with only a light sentence for theft because the judge ruled that in the light of Scott's new findings robots came under human law and therefore no infraction of justice had been committed. Working independently in his own laboratory Scott had proved that the magnetic flux lines in male and female robot systems, while at first deteriorating to both, were actually behaving according to the para-emotional theories of von Bohler. Scott termed the condition 'hysteric puppy-love' which, he claimed, had many of the advantages of human love if allowed to develop freely. Well, neither Min nor I pretended we understood all his equations but they sure made a stir among the scientists.

Frank kept getting more and more publicity. First we heard he was serving his sentence in the mech correction center at La Jolla, then we got a report that he'd turned up in Hollywood. Later it came out that Galact-A-vision Pictures had hired Frank for a film and had gone $10,000 bail for him. Not long after that he was getting billed all over Terra as *the* sensational first robot star.

All during the production of *Forbidden Robot Love* Frank remained lead copy for the newspapers. Reporters liked to write him up as the Valentino of the Robots. Frank

Nineteen Fan Clubs, usually formed by lonely female robots against their employers' wishes, sprang up spontaneously through the East and Middle West. Then somebody found out Frank could sing and the human teen-agers began to go for him. It got so everywhere you looked and everything you read, there was Frank staring you in the face. Frank in tweeds on the golf course. Frank at Ciro's or the Brown Derby in evening clothes. Frank posing in his sports jetabout against a blue Pacific background.

Meanwhile everybody forgot about Elizabeth Seven. The movie producers had talked about hiring her as Frank's leading lady until they found out about a new line of female robots that had just gone on the market. When they screen-tested the whole series and picked a lovely Mylar rationaloid named Diana Twelve, it hit Elizabeth pretty hard. She began to let herself go after that and Min and I didn't have the heart to say anything to her. It was pretty obvious she wasn't oiling herself properly, her hair wasn't brushed and she didn't seem to care when one of her photons went dead.

When *Forbidden Robot Love* premiered simultaneously in Hollywood and New York the critics all gave it rave reviews. There were pictures of Diana Twelve and Frank making guest appearances all over the country. Back at the Io we got in the habit of letting Elizabeth watch TV with us sometimes in the Renting Office and one night there happened to be an interview with Frank and Diana at the Sands Hotel in Las Vegas. I guess seeing the pretty robot starlet and her Frank sitting so close together in the nightclub must have made the girl mech feel pretty bad. Even then she didn't say a word against the big servo; she just never watched the set again after that.

When we tabbed up the Io's receipts that year they were so good Min and I decided to take a month off for an Earthside vacation. Min's retired brother in Berkeley was nice enough to come out and look after the place for us while we spent four solid weeks soaking up the sun in Southern California. When we got back out to the spotel, though, I could see there was something wrong by the look on Jim's face.

"It's that girl robot of yours, Bill," he said. "She's gone and deactivated herself."

We went right to 22A and found Elizabeth Seven stretched out on the floor. There was a screwdriver clutched in her hand and the relay banks in her side were exposed and horribly blackened.

"Crazy mech shorted out her own DX," Jim said.

Min and I knew why. After Jim left for Earth we dismantled Elizabeth the best we could and put her back in Frank's old locker. We didn't know what else to do with her.

Anyway, the slack season came and went and before long we were doing the spring cleaning again and wondering how heavy the Jovian Moons trade was going to be. I remember I'd been making some repairs outside and was just hanging up my spacesuit in the Renting Office when I heard the radar announcing a ship.

It was the biggest Marvin 990 I'd ever seen that finally suctioned up to the boom and secured. I couldn't take my eyes off the ship. She was pretty near the last word in rockets and loaded with accessories. It took me a minute or two before I noticed all the faces looking out of the viewports.

"Min!" I whispered. "There's something funny about those faces. They look like—"

"Robots!" Min answered. "Bill, that 990 is full of mechs!"

Just as she said it a bulky figure in white space fatigues swung out of the hatch and hurried up the gangplank. Seconds later it burst through the airlock.

"Frank Nineteen!" we gasped together.

"Please, where is Elizabeth?" he hummed anxiously. "Is she all right? I have to know."

Frank stood perfectly still when I told him about Elizabeth's self-deactivation; then a pitiful shudder went through him and he covered his face with his big Neoprene hands.

"I was afraid of that," he said barely audibly. "Where—you haven't—?"

"No," I said. "She's where you always kept her."

With that the big servo-pilot took off for 22A like a berserk robot and we were right behind him. We watched him tear open his old locker and gently lay out the girl's mech's parts so he could study them. After a minute or two he gave a long sigh and said, "Fortunately it's not as bad as I thought. I believe I can fix her." Frank worked hard over the blackened relays for twenty minutes, then he set the unit aside and began assembling the girl. When the final connections were made and the damaged unit installed he flicked on her power. We waited and nothing happened.

Five minutes went by. Ten. Slowly the big robot turned away, his broad shoulders drooping slightly.

"I've failed," he said quietly. "Her DX doesn't respond to the gain."

The girl mech, in her blue dress, lay there motionless where Frank had been working on her as the servo-pilot muttered over and over, "It's my fault, I did this to you."

Then Min shouted: "Wait! I heard something!"

There was a slow click of a relay—and movement. Painfully Elizabeth Seven rose on one elbow and looked around her.

"Frank, darling," she murmured, shaking her head. "I know you're just old memory tape. It's all I have left."

"Elizabeth, it's really me! I've come to take you away. We're going to be together from now on."

"*You*, Frank? This isn't just old feedback? You've come back to me?"

"Forever, darling. Elizabeth, do you remember what I said about those wonderful green little worlds, the asteroids? Darling, we're *going* to one of them! You and the others will love Alinda, I know you will. I've been there many times."

"Frank, is your DX all right? What *are* you talking about?"

"How stupid of me, darling—you haven't heard. Elizabeth, thanks to Dr. Scott, Congress has passed Robot Civil

Rights! And that movie I made helped swing public opinion to our side. We're free!

"The minute I heard the news I applied to Interplanetary for homestead rights on Alinda. I made arrangements to buy a ship with the money I'd earned and then I put ads in all the Robot Wanted columns for volunteer colonizers. You should have seen the response! We've got thirty robot couples aboard now and more coming later. Darling, we're the first pioneer wave of free robots. On board we have tons of supplies and parts—everything we need for building a sound robot culture."

"Frank Nineteen!" said the girl mech suddenly. "I should be furious with you. You and that Diana Twelve—I thought—"

The big servo gave a flat whirring laugh. "Diana and me? But that was all publicity, darling. Why, right at the start of the filming Diana fell in love with Sam Seventeen, one of the other actors. They're on board now."

"Robot civilization," murmured the girl after a minute. "Oh, Frank, that means robot government, robot art, robot science ..."

"And robot marriage," hummed Frank softly. "There has to be robot law, too. I've thought it all out. As skipper of the first robot-owned rocket, I'm entitled to marry couples in deep space at their request."

"But who marries us, darling? You can't do it yourself."

"I thought of that, too," said Frank, turning to me. "This human gentleman has every right to marry us. He's in command of a moving body in space just like the captain of a ship. It's perfectly legal, I looked it up in the Articles of Space. Will you do it, sir?"

Well, what could I say when Frank dug into his fatigues and handed me a Gideon prayer book marked at the marriage service?

Elizabeth and Frank said their I do's right there in the Renting Office while the other robot colonizers looked on. Maybe it was the way I read the service. Maybe I should have been a preacher, I don't know. Anyway, when I pronounced Elizabeth and Frank robot and wife, that whole bunch of lovesick mechs wanted me to do the job for them, too. Big copper work robots, small aluminum sales-girl mechs, plastoid clerks and typists, squatty little Mumetal lab servos, rationaloids, non-rationaloids and just plain sub-robots—all sizes and shapes. They all wanted individual ceremonies, too. It took till noon the next day before the last couple was hitched and the 990 left for Alinda.

Like I said, the spotel business isn't so different from the motel game back in California. Sure, you got improvements to make but a new sideline can get to be pretty profitable—if you get in on the ground floor.

Min and I got to thinking of all those robot colonizers who'd be coming out here. Interplanetary cleared the license just last week. Min framed it herself and hung it next to our orbit license in the Renting Office. She says a lot of motel owners do all right as Justices of the Peace.

Benefactor

by George H. Smith

This story is taken from *Fantastic Universe,* August 1958.

We can anticipate that robots will be fiercely resented, at first, in a society that will see them as the latest—and an indestructible—widespread threat to the workers whom they will replace. The men who will seek to alter the status quo will be called "robot lovers" and stoned. But what happens next?

He clutched at the lever with more force than he'd intended. It was set for further in the future....

"They're crazy! They're insane! That mob outside is made up of madmen," Jacob Clark told his young assistant, Bill Towney.

"They'll be battering at the door any minute now, sir," Towney said nervously.

"But why? Why are they doing it? My inventions have advanced the world a hundred years. I've always been a benefactor of man, not a destroyer."

"It's the robots. People are in a rage because they say the robots cause unemployment by replacing workers."

"It's utter nonsense, you know," Clark said impatiently. "Why can't they see that my intelligent, self-controlled robots are the greatest boon the human race has ever received from one man?"

"I don't know, sir, but they don't." Towney paused as the shouting and pounding outside became more intense. "They demanded that you take the robots out of the labor market and order your factories to stop making them. This is the result of your refusal."

"DOWN WITH CLARK! DOWN WITH THE TIN MEN! DEATH TO THE ROBOT LOVERS!" The furious mob was battering at the door now.

"Really, sir," Towney said, "you should leave here. They'll kill you if you don't!"

"Leave here? I should say not. I'll defy the fools. I'll tell them what I've done for them and make them understand." He glanced nervously at the door. "Besides there's only one door. I couldn't get away now."

"There's the time machine, sir."

"But isn't there some other way? Perhaps if you went out and talked to them...."

"You know there's no other way. Those people believe you've brought disaster to the human race and they mean to kill you. And if you don't hurry they will," Towney said urgently. "The time machine is set for twenty years in the future. Please hurry, sir!"

The door was beginning to give. Clark looked around unhappily and then walked to the time machine. "All right, I'll go. In the future I know the results of my work will be appreciated. I'll be a hero and benefactor of mankind."

Towney heard the door crash and roughly pushed his employer into the time machine as the mob burst through. "Push the starting button, push the starting button. Quick!" he screamed as the first of the mob reached him.

Clark's hand leaped to the control lever just as a brick crashed into his head. His hand completed its motion with more force than he had intended as he sank unconscious to the floor and the machine was set for a thousand years in the future instead of twenty.

The year three thousand had been a brilliant one for the robots, in fact, the most brilliant since the last human being had died some five hundred years before. They had reached Venus and Mars and were now planning a trip to Jupiter. And this very day, a huge statue of Jacob Clark, the creator and benefactor, was to be dedicated on the site of what once had been his laboratory. It seemed a shame that most of the records concerning him and his time had been lost in one of the great wars that had helped to extinguish the humans. The statue though was good for surely he looked like a robot. One of the few human books still in existence said that the Creator had created in his own image.

It was right at the foot of his own statue that the Guardians of the Shrine found Jacob Clark. They picked up his unconscious, bleeding body and laid it tenderly on a nearby bench. They bent over him with all the gentleness and solicitude that had been installed in his very first models and had been handed down from generation to generation of robots. They wanted to help him but they were very puzzled.

"Perhaps it came from a far part of the earth," one of them said.

"Or maybe a mistake was made at one of the birth factories," said another. "See, it is losing oil at a great rate."

"Perhaps," mused the elder, "it is a new model. At any rate it is a robot and has been damaged. As our great creator taught us, he must be aided. We will take it to the central repair factory in the city."

"But," the first robot protested, "it's awfully bulky to be carried so far."

Being creatures of logic, they thought about it for a moment and then the elder came to a decision that was both effective and reasonable.

"Since he is so bulky, we will disassemble him for transportation purposes," he said as he leaned over and gently twisted off Clark's right arm.

"Rather primitive and messy construction, I'd say," said the second robot as he tenderly unscrewed Jacob Clark's head from his body.

The Stutterer

by R.R. Merliss

Illustrated by Riley

This story is taken from *Astounding Science Fiction,* April 1955.

A man can be killed by a toy gun—he can die of fright, for heart attacks can kill. What, then, is the deadly thing that

must be sealed away, forever locked in buried concrete—a thing or an idea?

Out of the twenty only one managed to escape the planet. And he did it very simply, merely by walking up to the crowded ticket window at one of the rocket ports and buying passage to Earth. His Army identification papers passed the harassed inspection of the agent, and he gratefully and silently pocketed the small plastic stub that was handed him in exchange for his money.

He picked his way with infinite care through the hordes of ex-soldiers clamoring for passage back to the multitudinous planets from which they had come. Then he slowly climbed the heavy ramp into the waiting rocket.

He saw with relief that the seats were strongly constructed, built to survive the pressure of many gravities and he chose one as far removed as possible from the other passengers.

He was still very apprehensive, and, as he waited for the rocket to take off, he tried hard to remember the principles of the pulse drive that powered the ship, and whether his additional weight would upset its efficiency enough to awaken suspicion.

The seats filled quickly with excited hurrying passengers. Soon he heard the great door clang shut, and saw the red light flicker on, warning of the take-off. He felt a slow surge of pressure as the ship arose from the ground, and his chair creaked ominously with the extra weight. He became fearful that it might collapse, and he strained forward trying to shift some of the pressure through his feet to the floor. He sat that way, tense and immobile, for what seemed a long time until abruptly the strain was relieved and he heard the rising and falling whine of the rockets that told him the ship was in pulse drive, flickering back and forth across the speed of light.

He realized that the pilots had not discovered his extra weight, and that the initial hazards were over. The important thing was to look like a passenger, a returning soldier like the others, so that no one would notice him and remember his presence.

His fellow travelers were by this time chatting with one another, some playing cards, and others watching the teledepth screens. These were the adventurers who had flocked from all corners of the galaxy to fight in the first national war in centuries. They were the uncivilized few who had read about battle and armed struggle in their history books and found the old stories exciting.

They paid no attention to their silent companion who sat quietly looking through the quartz windows at the diamond-bright stars, tacked against the blackness of infinity.

The fugitive scarcely moved the entire time of the passage. Finally when Earth hung out in the sky like a blue balloon, the ship cut its pulsations and swung around for a tail landing.

The atmosphere screamed through the fins of the rocket, and the continents and the countries, and then the rivers and the mountains took shape. The big ship settled down as gently as a snowflake, shuddered a few times and was quiet.

The passengers hurriedly gathered up their scattered belongings and pushed toward the exit in a great rush to be out and back on Earth.

The fugitive was the last to leave. He stayed well away from the others, being fearful that, if he should touch or brush up against someone, his identity might be recognized.

When he saw the ramp running from the ship to the ground, he was dismayed. It seemed a flimsy structure, supported only by tubular steel. Five people were walking down it, and he made a mental calculation of their weight—about eight hundred pounds he thought. He weighed five times that. The ramp was obviously never built to support such a load.

He hesitated, and then he realized that he had caught the eye of the stewardess waiting on the ground. A little panicky, he stepped out with one foot and he was horrified to feel the steel buckle. He drew back hastily and threw a quick glance at the stewardess. Fortunately at the moment she was looking down one field and waving at someone.

The ramp floor was supported by steel tubes at its edges and in its exact center. He tentatively put one foot in the middle over the support and gradually shifted his weight to it. The metal complained creakily, but held, and he slowly trod the exact center line to Earth. The stewardess' back was turned toward him as he walked off across the field toward the customhouse.

He found it comforting to have under his feet what felt like at least one yard of cement. He could step briskly and not be fearful of betraying himself.

There was one further danger: the customs inspector.

He took his place at the end of the line and waited patiently until it led him up to a desk at which a uniformed man sat, busily checking and stamping declarations and traveling papers. The official, however, did not even look up when he handed him his passport and identification.

"Human. You don't have to go through immigration," the agent said. "Do you have anything to declare?"

"N-no," the traveler said. "I d-didn't bring anything in."

"Sign the affidavit," the agent said and pushed a sheet of paper toward him.

The traveler picked up a pen from the desk and signed "Jon Hall" in a clear, perfect script.

The agent gave it a passing glance and tossed it into a wire basket.

Then he pushed his uniform cap back exposing a bald head. "You're my last customer for a while, until the rocket from Sirius comes in. Guess I might as well relax for a minute." He reached into a drawer of the desk and pulled out a package of cigarettes, of which he lit one.

"You been in the war, too?" he asked.

Hall nodded. He did not want to talk any more than he had to.

The agent studied his face.

"That's funny," he said after a minute. "I never would have picked you for one of these so-called adventurers. You're too quiet and peaceful looking. I would have put you down as a doctor or maybe a writer."

"N-no," Hall said. "I w-was in the war."

"Well, that shows you can't tell by looking at a fellow," the agent said philosophically. He handed Hall his papers. "There you are. The left door leads out to the copter field. Good luck on Earth!"

Hall pocketed the stamped documents. "Thanks," he said. "I'm glad to be here."

He walked down the wide station room to a far exit and pushed the door open. A few steps farther and he was standing on a cement path dug into a hillside.

Across the valley, bright in the noon sun lay the pine covered slopes of the Argus mountains, and at his feet the green Mojave flowering with orchards stretched far to the north and south. Between the trees, in the center of the valley, the Sacramento River rolled southward in a man-made bed of concrete and steel giving water and life to what had a century before been dry dead earth.

There was a small outcropping of limestone near the cement walk, and he stepped over to it and sat down. He would have been happy to rest and enjoy for a few moments his escape and his triumph, but he had to let the others know so that they might have hope.

He closed his eyes and groped across the stars toward Grismet. Almost immediately he felt an impatient tug at his mind, strong because there were many clamoring at once to be heard. He counted them. There were seventeen. So one more had been captured since he had left Grismet.

"Be quiet," the told them. "I'll let you see, after a while. First I have to reach the two of us that are still free."

Obediently, the seventeen were still, and he groped some more and found another of his kind deep in an ice cave in the polar regions of Grismet.

"How goes it?" he asked.

The figure on Grismet lay stretched out at full length on the blue ice, his eyes closed. He answered without moving: "They discovered my radiation about an hour ago. Pretty soon, they'll start blasting through the ice."

The one on Earth felt the chill despair of his comrade and let go. He groped about again until he found the last one, the only other one left. He was squatting in the cellar of a warehouse in the main city of Grismet.

"Have they picked up your trail yet?" he asked.

"No," answered the one in the cellar. "They won't for a while. I've scattered depots of radiation all through the town. They'll be some time tracking them all down, before they can get to me."

In a flash of his mind, Hall revealed his escape and the one on Grismet nodded and said: "Be careful. Be very careful. You are our only hope."

Hall returned then to the seventeen, and he said with his thoughts: "All right, now you can look." Immobile in their darkness, they snatched at his mind, and as he opened his eyes, they, too, saw the splendors of the mountains and the valley, the blue sky, and the gold sun high overhead.

The new man was young, only twenty-six. He was lean and dark and very enthusiastic about his work. He sat straight in his chair waiting attentively while his superior across the desk leafed through a folder.

"Jordan. Tom Jordan," the older man finally said. "A nice old Earth name. I suppose your folks came from there."

"Yes, sir," the new man said briskly.

The chief closed the folder.

"Well," he said, "your first job is a pretty important one."

"I realize that, sir," Jordan said. "I know it's a great responsibility for a man just starting with the Commission, but I'll give it every thing I have."

The chief leaned back in his seat and scratched his chin thoughtfully.

"Normally we start a beginner like you working in a pair with an older man. But we just haven't got enough men to go around. There are eight thousand planets there"—he pointed with his thumb over his shoulder to a wall-sized map of the galaxy—"and we've got to cover every one. It seems reasonable that if he escaped this planet, he'll go to another that will by its atmosphere or its temperature give him some natural advantage over us—some place that is either burning hot or at absolute zero, or perhaps with a chlorine or sulfur dioxide atmosphere. That's why"—he hesitated a minute, but continued because he was a truthful man—"I picked you for Earth. It's the most populated of all the planets and it seems the least likely one that he would choose."

Jordan's face dropped a little bit when he heard the last piece of information, but he said: "I understand, sir, and if he's there, I'll bring him back."

The chief slouched farther back in his seat. He picked up a shard of rubidium that served as a paper weight and toyed with it.

"I guess you know most of the facts. They are made out of permallium. Have you ever seen any of the stuff?"

The new man shook his head. "I read about it though—some new alloy, isn't it?"

"Plenty new. It's the hardest stuff anybody has ever made. If you set off one hundred successive atom blasts over a lump of permallium, you might crystallize and scale maybe a micron off the surface. It will stand any temperature or pressure we can produce. That just means there's no way to destroy it."

Jordan nodded. He felt a little honored that the chief was giving him this explanation in person rather than just turning him over to one of the scientific personnel for a briefing. He did not understand that the old man was troubled and was talking the situation through as much for his own sake as for anyone else's.

"That's the problem," the chief continued. "Essentially an indestructible machine with a built-in source of power that one can't reach. It had to be built that way—a war instrument, you know."

He stopped and looked squarely at the bright young man sitting across the desk. "This lousy war. You'd think the human race would grow up some time, wouldn't you?" He filled a pipe with imported Earth tobacco and lit it, and took a few deep puffs. "There's something else. I don't know how they do it, but they can communicate with one another over long distances. That made them very useful for military purposes.

"They are loyal to one another, too. They try to protect each other and keep one another from being captured. Do you find that surprising?"

The question caught Jordan unprepared. "Well, yes. It is, kind of—" he said. "They are only machines."

The chief closed his eyes for a moment. He seemed tired.

"Yes," he repeated, "they are only machines. Anyway, we don't know everything about them, even yet. There are still a few secret angles, I think. The men who could tell us are either dead or in hiding.

"There's one fact though that gives us a great advantage. Their brain"—he stopped on the word and considered it— "I mean their thinking apparatus gives off a very penetrating short-wave length radiation which you can pick up on your meters anywhere in a radius of two thousand miles, and you can locate the source accurately if you get within fifty miles.

"The only real problem you'll have in finding them is the confusion created by illegal atomic piles. You'd be surprised how many of them we have turned up recently. They are owned by private parties and are run illegally to keep from paying the tax on sources of power. You have to track those down, but once you get them labeled it will be clear sailing."

He stopped to take a few puffs on his pipe.

"Don't try to be a hero," he said after a few moments. "Don't get close to the thing you are hunting. None of them yet has injured any of us, but if one should want to, he could crush you to death with two fingers. Use the permallium nets and net bombs if you locate him."

He tamped his pipe out. "Well, that's it," he said.

The new man arose. "I want you to know that I appreciate the trust you have put in me."

"Sure, sure," the chief said, but it was not unfriendly. "Do you like the job?"

"It is a great opportunity," Jordan said, and he meant it.

"What do you think about what we do to them after we capture them?"

The new man shrugged. "I suppose it's the only thing to do. It's not as though they were human."

"Yeah," the chief said. "I guess so. Anyway, good luck."

Jordan arose and shook the chief's hand. However, just as he was stepping through the door, his superior asked him another question. "Did you know that one of them stutters?"

He turned back, puzzled. "Stutters? Why should he stutter? How could that be?"

The chief shook his head and started cleaning out his pipe.

"I don't know for sure. You'd better get started." He sat back in his seat and watched the back of the new man as he disappeared through the doorway.

That young fellow has a lot to learn, he thought to himself. But even so, maybe he's better off than I am. Maybe I've had too much experience. Maybe too much experience puts you back where you started from. You've done the wrong thing so many times and profited so many times from your mistakes that you see errors and tragedies in everything.

He was depressed, and he did something that usually made him feel better again. He reached under the edge of his desk and pulled a little switch that made the galactic map on the wall light up in three-dimensional depth, then he swung around in his chair so he could see it. Eight thousand planets that his race had conquered, eight thousand planets hundreds of light-years apart. Looking at the map gave him a sense of accomplishment and pride in humanity which even a stupid war and its aftermath could not completely destroy.

Jon Hall, the fugitive, walked along the highway leading south from the rocket port. There was very little traffic, only an occasional delivery truck carrying meat or groceries. The real highway was half a mile overhead where the copters shuttled back and forth up and down the state in neat orderly layers.

The seventeen were inside his head, looking through his eyes, and feasting on the blueness of the sky, and the rich green vegetation that covered the fertile fields. From time to time they talked to him, giving advice, asking questions, or making comments, but mostly they looked, each knowing that the hours of their sight might be very few.

After walking a while, Hall became aware of someone's footsteps behind him. He stopped suddenly in apprehension and swung around. A dozen or so paces away was a red-headed boy of about ten or eleven, dressed in plastic overalls, and carrying a basket of ripe raspberries. The stains about his mouth showed that not all the raspberries were carried in the basket.

Hall's anxiety faded, and he was glad to see the child. He had hoped to meet someone who was not so old that they would become suspicious, but old enough that they might give him directions.

He waited for the lad to catch up.

"Hello," the boy said. "I've been walking behind you most of a mile, but I guess you didn't hear me."

"It looks as though you've been p-p-picking raspberries," Hall said.

"Yup. My dad owns a patch by the river. Want some?" He proffered the basket.

"No, thank you," Hall answered. He resumed his walk up the highway with the boy at his side.

"D-do you live around here," he asked.

"Just up the road a ways." The lad studied his companion for a minute. "You stutter, don't you?"

"A little."

"There was a boy in my class who used to stutter. The teacher said it was because he thought so far ahead of what he said he got all tangled up." The boy reached in his basket for a handful of berries and chewed them thoughtfully. "She was always after him to talk slower, but I guess it didn't do any good. He still stutters."

"Is there a p-power plant around here?" Hall asked. "You know, where the electricity comes from."

"You mean the place where they have the nu-nuclear fission"—the boy stumbled on the unfamiliar word, but got it out—"and they don't let you in because you get poisoned or something?"

"Yes, I think that's it."

"There are two places. There's one over at Red Mountain and another over at Ballarat."

"Where are they?"

"Well—" The boy stopped to think. "Red Mountain's straight ahead, maybe ten miles, and Ballarat's over there"—he pointed west across the orange groves—"maybe fifteen miles."

"Good," Hall said. "Good." And he felt glad inside of himself. Maybe it could be done, he thought.

They walked along together. Hall sometimes listening to the chattering of the boy beside him, sometimes listening to and answering the distant voices of the seventeen. Abruptly, a few hundred yards before the house that the boy had pointed out as his father's, a small sports car whipped down the highway, coming on them almost without warning. The lad jumped sideways, and Hall, to avoid touching him, stepped off the concrete road. His leg sank into the earth up to the mid-calf. He pulled it out as quickly as he could.

The boy was looking at the fast retreating rear of the sports car.

"Gee," he said. "I sure didn't see them coming." Then he caught sight of the deep hole alongside the road, and he stared at it. "Gosh, you sure made a footprint there," he said wonderingly.

"The ground was soft," Hall said. "C-come along."

But instead of following, the boy walked over to the edge of the road and stared into the hole. He tentatively stamped on the earth around it. "This ground isn't soft," he said. "It's hard as a rock." He turned and looked at Hall with big eyes.

Hall came close to the boy and took hold of his jacket. "D-don't pay any attention to it, son. I just stepped into a soft spot."

The boy tried to pull away. "I know who you are," he said. "I heard about you on the teledepth."

Suddenly, in the way of children, panic engulfed him and he flung his basket away and threw himself back and forth, trying to tear free. "Let me go," he screamed. "Let me go. Let me go."

"Just l-listen to me, son," Hall pleaded. "Just listen to me. I won't hurt you."

But the boy was beyond reasoning. Terror stricken, he screamed at the top of his voice, using all his little strength to escape.

"If you p-promise to l-listen to me, I'll let you go," Hall said.

"I promise," the boy sobbed, still struggling.

But the moment Hall let go of his coat, he tore away and ran as fast as he could over the adjacent field.

"W-wait—don't run away," Hall shouted. "I won't hurt you. Stay where you are. I couldn't follow you anyway. I'd sink to my hips."

The logic of the last sentence appealed to the frightened lad. He hesitated and then stopped and turned around, a hundred feet or so from the highway.

"L-listen," said Hall earnestly. "The teledepths are wr-wrong. They d-didn't tell you the t-truth about us. I d-don't want to hurt anyone. All I n-need is a few hours. D-don't tell anyone for j-just a few hours and it'll be all right." He paused because he didn't know what to say next.

The boy, now that he seemed secure from danger had recovered his wits. He plucked a blade of grass from the ground and chewed on an end of it, looking for all the world like a grownup farmer thoughtfully considering his fields. "Well, I guess you could have hurt me plenty, but you didn't," he said. "That's something."

"Just a few hours," Hall said. "It won't take long. Y-you can tell your father tonight."

The boy suddenly remembered his raspberries when he saw his basket and its spilled contents on the highway.

"Why don't you go along a bit," he said. "I would like to pick up those berries I dropped."

"Remember," Hall said, "just a few hours." He turned and started walking again toward Red Mountain. Inside his mind, the seventeen asked anxiously, "Do you think he'll give the alarm? Will he report your presence?"

Back on the highway, the boy was gathering the berries back into his basket while he tried to make his mind up.

Jordan reached Earth atmosphere about two o'clock in the afternoon. He immediately reported in to the Terrestrial police force, and via the teledepth screen spoke with a bored lieutenant. The lieutenant, after listening to Jordan's account of his mission, assured him without any particular enthusiasm of the willingness of the Terrestrial forces to coöperate, and of more value, gave him the location of all licensed sources of radiation in the western hemisphere.

The galactic agent set eagerly to work, and in the next several hours uncovered two unlisted radiation sources, both of which he promptly investigated. In one case, north of Eugene, he found in the backyard of a metal die company a small atomic pile. The owner was using it as an illegal generator of electricity, and when he saw Jordan snooping about with his detection instruments, he immediately offered the agent a sizable bribe. It was a grave mistake since Jordan filed charges against him, via teledepth, not only for evading taxes, but also for attempted bribery.

The second strike seemed more hopeful. He picked up strong radiation in a rather barren area of Montana; however when he landed, he found that it was arising from the earth itself. From a short conversation with the local authorities, he learned that the phenomenon was well known: an atomic fission plant had been destroyed at that site during the Third World War.

He was flying over the lovely blue water of Lake Bonneville, when his teledepth screen flickered. He flipped the switch on and the lieutenant's picture flooded in.

"I have a call I think you ought to take," the Earth official said. "It seems as though it might be in your line. It's from a sheriff in a small town in California. I'll have the operator plug him in."

Abruptly the picture switched to that of a stout red-faced man wearing the brown uniform of a county peace officer.

"You're the galactic man?" the sheriff asked.

"Yes. My name is Tom Jordan," Jordan said.

"Mine's Berkhammer." It must have been warm in California because the sheriff pulled out a large handkerchief and mopped his brow. When he was done with that he blew his nose loudly. "Hay fever," he announced.

"Want to see my credentials?"

"Oh sure, sure," the sheriff hastily replied. He scrutinized the card and badge that Jordan displayed. After a moment, he said, "I don't know why I'm looking at those. They might be fakes for all I know. Never saw them before and I'll probably never see them again."

"They're genuine."

"The deuce with formality," the sheriff said heavily. "There's some kid around here who thinks he saw that ... that machine you're supposed to be looking for."

"When was that?" Jordan asked.

"About four hours ago. Here, I'll let you talk to him yourself." He pulled his big bulk to one side, and a boy and his father walked into the picture. The boy was red-eyed, as though he had been crying. The father was a tall, stoop-shouldered farmer, dressed like his son in plastic overalls.

The sheriff patted the boy on the back. "Come on, Jimmy. Tell the man what you saw."

"I saw him," the boy said sullenly. "I walked up the highway with him."

Jordan leaned forward toward the screen.

"How did you know who he was?"

"I knew because when he stepped on the ground, he sank into it up to his knee. He tried to say the ground was soft, but it was hard. I know it was hard."

"Why did you wait so long to tell anybody?" Jordan asked softly.

The boy looked at him with defiance and dislike in his eyes and kept his small mouth clamped shut.

His father nudged him roughly in the ribs.

"Answer the man," he commanded.

Jimmy looked down at his shoes.

"Because he asked me not to tell for a while," he said curtly.

"Stubborn as nails," the father said not without pride in his voice. "Got more loyalty to a lousy machine than to the whole human race."

"Which way did he go, Jimmy?"

"Toward Red Mountain. I think maybe to the power house. He asked me where it was."

"What do you think he wants with that?" the sheriff asked of Jordan.

Jordan shrugged and shook his head.

"Maybe it's all in the kid's head," the sheriff suggested. "These wild teledepth programs they look at give them all kinds of ideas."

"It isn't in my head," Jimmy said violently. "I saw him. He stepped on the ground and stuck his foot into it. I talked to him. And I know something else. He stutters."

"What?" said the sheriff. "Now I know you're lying."

The father started dragging the boy by the arm. "Come on home, Jimmy. You got one more licking coming."

Jordan, however, was sure the boy was not lying. "Leave him alone," he said. "He's right. He did see him." He took a fast look at the timepiece on his panel board. "I'll be down in an hour and a half. Wait for me."

He flicked the switch off, and kicked up the motors. The ship shot southward almost as rapidly as a projectile.

He had topped the Sierras and had just turned into the great central valley of California when, with the impact of a blow, a frightening thought occurred to him.

He flicked the screen on again, and he caught the sheriff sitting behind his desk industriously scratching himself in one armpit.

"Listen," Jordan said, speaking very fast. "You've got to send out a national alarm. You must get every man you can down to the power plant. You've got to stop him from getting in."

The sheriff stopped scratching himself and stared at Jordan.

"What are you so het up about, young man?"

"Do it, and do it now," Jordan almost shouted. "He'll tear the pile apart and let the hafnium go off. It'll blow half the state off the planet."

The sheriff was unperturbed. "Mr. Star boy," he said sarcastically, "any grammar school kid knows that if someone came within a hundred yards of one of those power-house piles, he'd burn like a match stick. And besides why would he want to blow himself to pieces?"

"He's made out of permallium." Jordan was shouting now.

The sheriff suddenly grew pale. "Get off my screen. I'm calling Sacramento."

Jordan set the ship for maximum speed, well beyond the safety limit. He kept peering ahead into the dusk, momentarily fearful that the whole countryside would light up in one brilliant flash. In a few minutes he was sweating and trembling with the tension.

Over Walnut Grove, he recognized the series of dams, reservoirs and water-lifts where the Sacramento was raised up out of its bed and turned south. For greater speed, he came close to Earth, flying at emergency height, reserved ordinarily for police, firemen, doctors and ambulances. He set his course by sight following the silver road of the river, losing it for ten or fifteen miles at a time where it passed through subterranean tunnels, picking it up again at the surface, always shooting south as fast as the atmosphere permitted.

At seven thirty, when the sun had finally set, he sighted the lights of Red Mountain, and he cut his speed and swung in to land. There was no trouble picking out the power plant; it was a big dome-shaped building surrounded by a high wall. It was so brilliantly lit up, that it stood out like a beacon, and there were several hundred men milling about before it.

He settled down on the lawn inside the walls, and the sheriff came bustling up, a little more red in the face than usual.

"I've been trying to figure for the last hour what the devil I would do to stop him if he decided to come here," Berkhammer said.

"He's not here then?"

The sheriff shook his head. "Not a sign of him. We've gone over the place three times."

Jordan settled back in relief, sitting down in the open doorway of his ship. "Good," he said wearily.

"Good!" the sheriff exploded. "I don't know whether I'd rather have him show up or not. If this whole business is nothing more than the crazy imagination of some kid who ought to get tanned and a star-cop with milk behind his ears, I'm really in the soup. I've sent out an alarm and I've got the whole state jumping. There's a full mechanized battalion of state troops waiting in there." He pointed toward the power plant. "They've got artillery and tanks all around the place."

Jordan jumped down out of the ship. "Let's see what you've got set up here. In the meantime, stop fretting. I'd rather see you fired than vaporized along with fifty million other people."

"I guess you're right there," Berkhammer conceded, "but I don't like to have anyone make a fool out of me."

At Ballarat, an old man, Eddie Yudovich, was the watchman and general caretaker of the electrical generation plant. Actually, his job was a completely unnecessary one, since the plant ran itself. In its very center, buried in a mine of graphite were the tubes of hafnium, from whose nuclear explosions flowed a river of electricity without the need of human thought or direction.

He had worked for the company for a long time and when he became crippled with arthritis, the directors gave him the job so that he might have security in his latter years.

Yudovich, however, was a proud old man, and he never once acknowledged to himself or to anyone else that his work was useless. He guarded and checked the plant as though it were the storehouse of the Terrestrial Treasury. Every hour punctually, he made his rounds through the building.

At approximately seven thirty he was making his usual circuit when he came to the second level. What he discovered justified all the years of punctilious discharge of his duties. He was startled to see a man kneeling on the floor, just above where the main power lines ran. He had torn a hole in the composition floor, and as Yudovich watched, he reached in and pulled out the great cable. Immediately the intruder glowed in the semidarkness with an unearthly blue shine and sparkles crackled off of his face, hands and feet.

Yudovich stood rooted to the floor. He knew very well that no man could touch that cable and live. But as he watched, the intruder handled it with impunity, pulling a length of wire out of his pocket and making some sort of a connection.

It was too much for the old man. Electricity was obviously being stolen. He roared out at the top of his voice, and stumped over to the wall where he threw the alarm switch. Immediately, a hundred arc lights flashed on, lighting the level brighter than the noon sun, and a tremendously loud siren started wailing its warning to the whole countryside.

The intruder jumped up as though he had been stabbed. He dropped the wires, and after a wild look around him, he ran at full speed toward the far exit.

"Hold on there," Yudovich shouted and tried to give chase, but his swollen, crooked knees almost collapsed with the effort.

His eyes fell on a large wrench lying on a worktable, and he snatched it up and threw it with all his strength. In his youth he had been a ball player with some local fame as a pitcher, and in his later life, he was addicted to playing horseshoes. His aim was, therefore, good, and the wrench sailed through the air striking the runner on the back of the head. Sparks flew and there was a loud metallic clang, the wrench rebounding high in the air. The man who was struck did not even turn his head, but continued his panicky flight and was gone in a second.

When he realized there was no hope of effecting a capture, Yudovich stumped over to see the amount of the damage. A hole had been torn in the floor, but the cable itself was intact.

Something strange caught his attention. Wherever the intruder had put his foot down, there were many radiating cracks in the composition floor, just as though someone had struck a sheet of ice with a sledge hammer.

"I'll be danged," he said to himself. "I'll be danged and double danged."

He turned off the alarm and then went downstairs to the teledepth screen to notify the sheriff's office.

A few hundred yards from the powerhouse, Jon Hall stood in the darkness, listening to the voices of his fellows. There were eighteen of them, not seventeen, for a short while before the one in the ice cave had been captured, and they railed at him with a bitter hopeless anger.

He looked toward the bright lights of the powerhouse, considering whether he should return. "It's too late," said one of them. "The alarm is already out." "Go into the town and mix with the people," another suggested. "If you stay within a half mile of the hafnium pile, the detection man will not be able to pick up your radiation and maybe you will have a second chance."

They all assented in that, and Hall, weary of making his own decisions turned toward the town. He walked through a tree-lined residential street, the houses with neatly trimmed lawns, and each with a copter parked on the roof. In almost every house the teledepths were turned on and he caught snatches of bulletins about himself: "... Is known to be in the Mojave area." "... About six feet in height and very similar to a human being. When last seen, he was dressed in—" "Governor Leibowitz has promised speedy action and attorney general Markle has stated—"

The main street of Ballarat was brilliantly lighted. Many of the residents, aroused by the alarm from the powerhouse, were out, standing in small groups in front of the stores and talking excitedly to one another.

He hesitated, unwilling to walk through the bright street, but uncertain where to turn. Two men talking loudly came around the corner suddenly and he stepped back into a store entrance to avoid them. They stopped directly in front of him. One of them, an overalled farm hand from his looks, said, "He killed a kid just a little while ago. My brother-in-law heard it."

"Murderer," the other said viciously.

The farmer turned his head and his glance fell on Hall. "Well, a new face in town," he said after a moment's inspection. "Say I bet you're a reporter from one of the papers, aren't you?"

Hall came out of the entrance and tried to walk around the two men, but the farmer caught him by the sleeve.

"A reporter, huh? Well, I got some news for you. That thing from Grismet just killed a kid."

Hall could restrain himself no longer.

"That's a lie," he said coldly.

The farmer looked him up and down.

"What do you know about it," he demanded. "My brother-in-law got it from somebody in the state guard."

"It's still a lie."

"Just because it's not on the teledepth, you say it's a lie," the farmer said belligerently. "Not everything is told on the teledepth, Mr. Wiseheimer. They're keeping it a secret. They don't want to scare the people."

Hall started to walk away, but the farmer blocked his path.

"Who are you anyway? Where do you live? I never saw you before," he said suspiciously.

"Aw, Randy," his companion said, "don't go suspecting everybody."

"I don't like anyone to call me a liar."

Hall stepped around the man in his path, and turned down the street. He was boiling inside with an almost uncontrollable fury.

A few feet away, catastrophe suddenly broke loose. A faulty section of the sidewalk split without warning under his feet and he went pitching forward into the street. He clutched desperately at the trunk of a tall palm tree, but with a loud snap, it broke, throwing him head on into a parked road car. The entire front end of the car collapsed like an egg shell under his weight.

For a long moment, the entire street was dead quiet. With difficulty, Hall pulled himself to his feet. Pale, astonished faces were staring at him from all sides.

Suddenly the farmer started screaming. "That's him. I knew it. That's him." He was jumping up and down with excitement.

Hall turned his back and walked in the other direction. The people in front of him faded away, leaving a clear path.

He had gone a dozen steps when a man with a huge double-barreled shotgun popped out from a store front just ahead. He aimed for the middle of Hall's chest and fired both barrels.

The blast and the shot struck Hall squarely, burning a large hole in his shirt front. He did not change his pace, but continued step by step.

The man with the gun snatched two shells out of his pocket and frantically tried to reload. Hall reached out and closed his hand over the barrel of the gun and the blue steel crumpled like wet paper.

From across the street, someone was shooting at him with a rifle. Several times a bullet smacked warmly against his head or his back.

He continued walking slowly up the street. At its far end several men appeared dragging a small howitzer—probably the only piece in the local armory. They scurried around it, trying to get it aimed and loaded.

"Fools. Stupid fools," Hall shouted at them.

The men could not seem to get the muzzle of the gun down, and when he was a dozen paces from it they took to their heels. He tore the heavy cannon off of its carriage and with one blow of his fist caved it in. He left it lying in the street broken and useless.

Almost as suddenly as it came, his anger left him. He stopped and looked back at the people cringing in the doorways.

"You poor, cruel fools," Hall said again.

He sat down in the middle of the street on the twisted howitzer barrel and buried his head in his hands. There was nothing else for him to do. He knew that in just a matter of seconds, the ships with their permallium nets and snares would be on him.

Since Jordan's ship was not large enough to transport Jon Hall's great weight back to Grismet, the terrestrial government put at the agent's disposal a much heavier vessel, one room of which had been hastily lined with permallium and outfitted as a prison cell. A pilot by the name of Wilkins went with the ship. He was a battered old veteran, given to cigar smoking, clandestine drinking and card playing.

The vessel took off, rose straight through the atmosphere for about forty miles, and then hung, idly circling Earth, awaiting clearance before launching into the pulse drive. A full course between Earth and Grismet had to be plotted and cleared by the technicians at the dispatch center because the mass of the vessel increased so greatly with its pulsating speed that if any two ships passed within a hundred thousand miles of each other, they would at least be torn from their course, and might even be totally destroyed.

Wilkins had proposed a pinochle game, and he and Jordan sat playing in the control room.

The pilot had been winning and he was elated. "Seventy-six dollars so far," he announced after some arithmetic. "The easiest day's pay I made this month."

Jordan shuffled the cards and dealt them out, three at a time. He was troubled by his own thoughts, and so preoccupied that he scarcely followed the game.

"Spades, again," the pilot commented gleefully. "Well, ain't that too bad for you." He gave his cigar a few chomps and played a card.

Jordan had been looking out of the window. The ship had tilted and he could see without rising the rim of Earth forming a beautiful geometric arc, hazy and blue in its shimmering atmosphere.

"Come on, play," the pilot said, impatiently. "I just led an ace."

Jordan put down his cards. "I guess I better quit," he said.

"What the devil!" the pilot said angrily. "You can't quit like that in the middle of a deal. I got a flush and aces."

"I'm sorry," Jordan said, "but I'm going to lie down in my cabin until we are given clearance."

He opened the door of the little room and went into the hall. He walked down past his own cabin and stopped in front of another door, a new one that was sheathed in permallium. He hesitated a few moments; then he snapped open the outside latch and walked in, letting the door swing closed behind him.

Hall lay unmoving in the middle of the floor, his legs and arms fastened in greaves of permallium.

Jordan was embarrassed. He did not look directly at the robot.

"I don't know whether you want to talk to me or not," he started. "If you don't want to, that's all right. But, I've followed you since you landed on Earth, and I don't understand why you did what you did. You don't have to tell me, but I wish you would. It would make me feel better."

The robot shrugged—a very human gesture, Jordan noted.

"G-go ahead and ask me," he said. "It d-doesn't make any difference now."

Jordan sat down on the floor. "The boy was the one who gave you away. If not for him, no one would have ever known what planet you were on. Why did you let the kid get away?"

The robot looked straight at the agent. "Would you kill a child?" he asked.

"No, of course not," Jordan said a little bit annoyed, "but I'm not a robot either." He waited for a further explanation, but when he saw none was coming, he said: "I don't know what you were trying to do in that powerhouse at Ballarat, but, whatever it was, that old man couldn't have stopped you. What happened?"

"I l-lost my head," the robot said quietly. "The alarm and the lights rattled me, and I got into a p-panic."

"I see," said Jordan, frustrated, not really seeing at all. He sat back and thought for a moment. "Let me put it this way. Why do you stutter?"

Hall smiled a wry smile. "Th-that used to be a m-military secret," he said. "It's our one weakness—the one Achilles heel in a m-machine that was meant to be invulnerable."

He struggled to a sitting position. "You see, we were m-made as s-soldiers and had to have a certain loyalty to the country that m-made us. Only living things are loyal—machines are not. We had to think like human beings."

Jordan's brows contracted as he tried to understand the robot.

"You mean you have a transplanted human brain?" he asked incredulously.

"In a way," Hall said. "Our b-brains are permallium strips on which the mind of some human donor was m-magnetically imprinted. My mind was copied f-from a man who stuttered and who got panicky when the going got rough, and who couldn't kill a child no matter what was at s-stake."

Jordan felt physically ill. Hall was human and he was immortal. And according to galactic decree, he, like his fellows, was to be manacled in permallium and fixed in a great block of cement, and that block was to be dropped into the deep silent depths of the Grismet ocean, to be slowly covered by the blue sediment that gradually filters down through the miles of ocean water to stay immobile and blind for countless millions of years.

Jordan arose to his feet. He could think of nothing further to say.

He stopped, however, with the door half open, and asked: "One more question—what did you want with the electrical generator plants on Earth?"

Slowly and without emotion Hall told him, and when he understood, he became even sicker.

He went across to his cabin and stood for a while looking out the window. Then he lit a cigarette and lay down on his bunk thinking. After a time, he put out the cigarette and walked into the hall where he paced up and down.

As he passed the cell door for about the tenth time, he suddenly swung around and lifted the latch and entered. He went over to the robot, and with a key that he took from his pocket, he unlocked the greaves and chains.

"There's no point in keeping you bound up like this," he said. "I don't think you're very dangerous." He put the key back in his pocket.

"I suppose you know that this ship runs on an atomic pile," he said in a conversational tone of voice. "The cables are just under the floor in the control room and they can be reached through a little trap door."

Jordan looked directly into Hall's face. The robot was listening with great intentness.

"Well," the agent said, "we'll probably be leaving Earth's atmosphere in about fifteen minutes. I think I'll go play pinochle with the pilot."

He carefully left the door of the cell unlatched as he left. He walked to the control room and found Wilkins, a dry cigar butt clenched between his teeth, absorbed in a magazine.

"Let's have another game," Jordan said. "I want some of that seventy-six dollars back."

Wilkins shook his head. "I'm in the middle of a good story here. Real sexy. I'll play you after we take off."

"Nothing doing," Jordan said sharply. "Let's play right now."

Wilkins kept reading. "We got an eighteen-hour flight in front of us. You have lots of time."

The agent snatched the magazine out of his hands. "We're going to play right now in my cabin," he said.

"You quit when I have aces and a flush, and now you come back and want to play again. That's not sportsmanlike," Wilkins complained, but he allowed himself to be led back to Jordan's cabin. "I never saw anybody so upset about losing a miserable seventy-six bucks," was his final comment.

The robot lay perfectly still until he heard the door to Jordan's cabin slam shut, and then he arose as quietly as he could and stole out into the hall. The steel of the hall floor groaned, but bore his weight, and carefully, trembling with excitement inside of his ponderous metallic body, he made his way to the control room. He knelt and lifted the little trap door and found the naked power cable, pulsating with electrical current.

In a locker under the panel board he found a length of copper wire. It was all he needed for the necessary connection.

Since his capture, his fellows on Grismet had been silent with despair, but as he knelt to close the circuit, their minds flooded in on him and he realized with a tremendous horror that there were now nineteen, that all except he had been bound and fixed in their eternal cement prisons.

"We are going to have our chance," he told them. "We won't have much time, but we will have our chance."

He closed the circuit and a tremendous tide of electric power flowed into his head. Inside that two-inch shell of permallium was a small strip of metal tape on whose electrons and atoms were written the borrowed mind of a man. Connected to the tape was a minute instrument for receiving and sending electromagnetic impulses—the chain by which the mind of one robot was tied to that of another.

The current surged in and the tiny impulses swelled in strength and poured out through the hull of the ship in a great cone that penetrated Earth's atmosphere in a quadrant that extended from Baffin land to Omaha, and from Hawaii to Labrador. The waves swept through skin and bone and entered the sluggish gelatinous brain of sentient beings, setting up in those organs the same thoughts and pictures that played among the electrons of the permallium strip that constituted Jon Hall's mind.

All nineteen clamored to be heard, for Hall to relay their voices to Earth, but he held them off and first he told his story.

The Casseiopeian delegate to the Galactic Senate was at the moment finishing his breakfast. He was small and furry, not unlike a very large squirrel, and he sat perched on a high chair eating salted roast almonds of which he was very fond.

Suddenly a voice started talking inside of his head, just as it did at that very second inside the heads of thirteen billion other inhabitants of the northwest corner of Earth. The Casseiopeian delegate was so startled that he dropped the dish of almonds, his mouth popping open, his tiny red tongue inside flickering nervously. He listened spellbound.

The voice told him of the war on Grismet and of the permallium constructed robots, and of the cement blocks. This, however, he already knew, because he had been one of the delegates to the Peace Conference who had decided to dispose of the robots. The voice, however, also told him things he did not know, such as the inability of the robots to commit any crime that any other sane human being would not commit, of their very simple desire to be allowed to live in peace, and most of all of their utter horror for the fate a civilized galaxy had decreed for them.

When the voice stopped, the Casseiopeian delegate was a greatly shaken little being.

Back on the ship, Hall opened the circuit to the nineteen, and they spoke in words, in memory pictures and in sensations.

A copter cab driver was hurrying with his fare from Manhattan to Oyster Bay. Suddenly, in his mind, he became a permallium robot. He was bound with cables of the heavy metal, and was suspended upside down in a huge cement block. The stone pressed firmly on his eyes, his ears, and his chest. He was completely immobile, and worst of all, he knew that above his head for six miles lay the great Grismet Ocean, with the blue mud slowly settling down encasing the cement in a stony stratum that would last till the planet broke apart.

The cab driver gasped: "What the hell." His throat was so dry he could scarcely talk. He turned around to his fare, and the passenger, a young man, was pale and trembling.

"You seeing things, too?" the driver asked.

"I sure am," the fare said unsteadily. "What a thing to do."

For fifteen minutes, over the northwest quadrant of Earth, the words and the pictures went out, and thirteen billion people knew suddenly what lay in the hearts and minds of nineteen robots.

A housewife in San Rafael was at the moment in a butcher shop buying meat for her family. As the thoughts and images started pouring into her mind, she remained stock-still, her package of meat forgotten on the counter. The butcher, wiping his bloodied hands on his apron froze in that position, an expression of horror and incredulity on his face.

When the thoughts stopped coming in, the butcher was the first to come out of the trancelike state.

"Boy," he said, "that's sure some way of sending messages. Sure beats the teledepths."

The housewife snatched her meat off the counter. "Is that all you think of," she demanded angrily.

"That's a terrible thing that those barbarians on Grismet are doing to those ... those people. Why didn't they tell us that they were human." She stalked out of the shop, not certain what she would do, but determined to do something.

In the ship Hall reluctantly broke off the connection and replaced the trap door. Then he went back to his cell and locked himself in. He had accomplished his mission; its results now lay in the opinions of men.

Jordan left the ship immediately on landing, and took a copter over to the agency building. His conversation with his superior was something he wanted to get over with as soon as possible.

The young woman at the secretary's desk looked at him coldly and led him directly into the inner office. The chief was standing up in front of the map of the galaxy, his hands in his pockets, his eyes an icy blue.

"I've been hearing about you," he said without a greeting.

Jordan sat down. He was tense and jumpy but tried not to show it. "I suppose you have," he said, adding, after a moment, "Sir."

"How did that robot manage to break out of his cell and get to the power source on the ship in the first place?"

"He didn't break out," Jordan said slowly. "I let him out."

"I see," the chief said, nodding. "You let him out. I see. No doubt you had your reasons."

"Yes, I did. Look—" Jordan wanted to explain, but he could not find the words. It would have been different if the robots' messages had reached Grismet; he would not have had to justify himself then. But they had not, and he could not find a way to tell this cold old man of what he had learned about the robots and their unity with men. "I did it because it was the only decent thing to do."

"I see," the chief said. "You did it because you have a heart." He leaned suddenly forward, both hands on his desk. "It's good for a man to have a heart and be compassionate. He's not worth anything if he isn't. But"—and he shook his finger at Jordan as he spoke—"that man is going to be compassionate at his own expense, not at the expense of the agency. Do you understand that?"

"I certainly do," Jordan answered, "but you have me wrong if you think I'm here to make excuses or to apologize. Now, if you will get on with my firing, sir, I'll go home and have my supper."

The chief looked at him for a long minute. "Don't you care about your position in the agency?" he asked quietly.

"Sure I do," Jordan said almost roughly. "It's the work I wanted to do all my life. But, as you said, what I did, I did at my own expense. Look, sir, I don't like this any better than you do. Why don't you fire me and let me go home? Your prisoner's safely locked up in the ship."

For answer the chief tossed him a stellogram. Jordan glanced at the first few words and saw that it was from Galactic Headquarters on Earth. He put it back on the desk without reading it through.

"I know that I must have kicked up a fuss. You don't have to spell it out for me."

"Read it," the chief said impatiently.

Jordan took back the stellogram and examined it. It read.

To: Captain Lawrence Macrae Detection Agency, Grismet.

From: Prantal Aminopterin Delegate from Casseiopeia Chairman, Grismet Peace Committee of the Galactic Senate.

Message: You are hereby notified that the committee by a vote of 17-0 has decided to rescind its order of January 18, 2214, directing the disposal of the permallium robots of Grismet. Instead, the committee directs that you remove from their confinement all the robots and put them in some safe place where they will be afforded reasonable and humane treatment.

The committee will arrive in Grismet some time during the next month to decide on permanent disposition.

Jordan's heart swelled as he read the gram. "It worked," he said. "They have changed their minds. It won't be so bad being discharged now." He put the paper back on the desk and arose to go.

The chief smiled and it was like sunlight suddenly flooding over an arctic glacier. "Discharged? Now who's discharging you? I'd sooner do without my right arm."

He reached in a desk drawer and pulled out a bottle of old Earth bourbon and two glasses. He carefully poured out a shot into each glass, and handed one to Jordan.

"I like a man with a heart, and if you get away with it, why then you get away with it. And that's just what you've done."

He sat down and started sipping his whisky. Jordan stood uncertainly above him, his glass in his hand.

"Sit down, son," the old man said. "Sit down and tell me about your adventures on Earth."

Jordan sat down, put his feet on the desk and took a sizable swallow of his whisky.

"Well, Larry," he started, "I got into Earth atmosphere about 2:40 o'clock—"

Robot Tales

www.ingramcontent.com/pod-product-compliance
Lightning Source LLC
LaVergne TN
LVHW090318210926
848480LV00001B/290